Kinetic Theory of Vehicular Traffic

Kinetic Theory of Vehicular Traffic

Ilya Prigogine
Université Libre de Bruxelles
Center for Statistical
Mechanics and Thermodynamics
University of Texas
Austin, Texas

Robert Herman
General Motors Research Laboratories
Warren, Michigan

American Elsevier
Publishing Company, Inc.
NEW YORK · 1971

AMERICAN ELSEVIER PUBLISHING COMPANY, INC.
52 Vanderbilt Avenue, New York, N.Y. 10017

ELSEVIER PUBLISHING COMPANY, LTD.
Barking, Essex, England

ELSEVIER PUBLISHING COMPANY
335 Jan Van Galenstraat, P.O. Box 211
Amsterdam, The Netherlands

International Standard Book Number 0-444-00082-8

Library of Congress Card Number 79-127772

Printed in the United States of America

Contents

Preface

The problems of transportation and traffic are very old ones. From time immemorial these problems have been deeply involved with the movements of man or the location of population concentrations. Throughout man's history transportation routes have shaped the character of the great migrations of peoples over all the continents. Such mass movements are not confined to the human family, since as we well know, birds, fishes, and various animals also undergo remarkable migrations, even on a yearly basis.

Transportation and traffic problems, in general, concern the interacting man–vehicle–road complex, and the rules of the game are spelled out, but only in part, by laws and social custom. Cities, the entities that satisfy man's need for clustering, have been crowded for a long time. In ancient Rome, where the population has been estimated to have reached about a half million, there were traffic problems exemplified by what we read in the *Corpus Inscriptionum Latinarum*: ". . . let no one after the Kalends of January drive or conduct a wagon between the time of sunrise and the tenth hour of the day along those streets which lead or will lead into the city of Rome in regions where there is populous habitation, unless it be necessary that material be carried or transported for the sake of building sacred temples of the immortal gods, or for the sake of public works, or unless it be necessary that materials that have been publicly contracted for demolition be carried out from the city or from these regions, on the condition that permission shall have been granted through this law to specific men in specific cases to drive or conduct wagons for this purpose."

We also recall with dismay the state of the road around London, especially in the seventeenth century during the time of the plague and the great fire, when narrow streets turned into quagmires by rain made the movement of people and their carts loaded with possessions virtually impossible. Toward the latter part of the last century we have photographic evidence of traffic jams in Philadelphia and New York City, for example, that stagger the imagination. These contributions to the growing social problems were made long before the motor vehicle. With the advent of the automobile there were many of the problems of today, including the waste of waiting—an everyday and all-day-long sight at the street crossings in any of our great cities, as well as the wailing of the motorists of one small community over the undue activity of the police and their taking measures to warn brother drivers by putting up a sign on an all but deserted street—"Beware of Motor Cops."

It goes without saying that during the present century great advances have been made in the development of the automobile as a very efficient means of personal transportation. This personalized transportation has given the vast majority of people in the United States, and the numbers are growing all over the world, a freedom that has changed the entire shape of our social structure. This would not have been possible but for the creation by our engineers of the transportation complex that has been so vitally responsible for the remarkable growth and development of our society.

It is unfortunate that this phenomenal burgeoning of traffic, associated with many personal and social ills, has not been matched by a corresponding increase in our fundamental understanding of vehicular traffic science. In fact, this area has unfortunately been developing essentially only over the past decade at a rather slow pace and has not yet emerged from an embryonic stage. In the past the chief concern has been with road building and town planning, even by such greats as Leonardo da Vinci, who wrote, "And know that if anyone wishes to go through the whole place by the high-level roads, he will be able to use them for this purpose, and so also if anyone wishes to go by the low-level roads. . . . The high-level roads are not to be used by wagons or vehicles such as these but are solely for the convenience of the gentlefolk. All carts and loads for the service and convenience of the common people should be confined to the low-level roads." Leonardo's fifteenth-century ideal city with two-level highways has provided much food for modern thought and was the keynote for the General Motors Futurama of 1939.

More recently, investigators in the physical sciences and applied mathematics have recognized that a very meaningful discipline lay virtually unexplored to this very day. There is no reason to believe that even so complex a subject as the driver–car–road problem in all its ramifications and implications is not amenable to a rational attack with the equipment of mathematics, theoretical and experimental physics, engineering, psychology, and medicine, to mention only a few of the varied disciplines required.

During recent years a mathematical approach, coupled with experiments and observations, has been applied to various, but unfortunately, far too few, specific traffic problems. Some of the problems investigated lie in the area of single-lane flow studied in terms of "follow-the-leader" theory, multiple-lane flow handled by stochastic or statistical theory, highway crossing and merging problems organized with queueing theory, and intersection and network problems. Long before broad systems knowledge can be synthesized, we feel that many specific narrow problems must be understood in order to furnish the fundamental understanding on which to build.

The purpose of the present work is to examine the features of a system in which each individual member, while trying to achieve his own optimum solution, has characteristics that depend on his being imbedded in a group. Of course, the characterization of the group then depends on the interaction between its members. The competition between individuals must, in the end, lead to either anarchy or some type of collective strategy. The latter is beautifully exemplified by the magnificently coordinated flight of a large flock of birds or the remarkable darting collective motion of a school of fish.

The analogy with economic theory is very striking. In the same way that no theory of economic behavior can be complete without an analysis of consumer behavior, no theory of vehicular traffic can ignore the problem of driver strategy. For this reason, traffic theory is a discipline that combines in a unique fashion concepts of mathematical statistics, physical theories, and human behavior.

We hope that the ideas presented in this monograph will encourage investigators interested in traffic to increase the level of the experimental effort in order to obtain the fundamental information required for a deeper understanding of this subject area. A large proportion of the material we are presenting here has appeared in our original publications cited in the appended list of references. We believe, however, that this systematic presentation provides a clearer development of the subject, permits the elucidation of a number of points, and provides the tools for further progress.

We acknowledge with pleasure the collaboration of the following scientific colleagues in the original work: Dr. R. L. Anderson, Professor F. C. Andrews, Professor R. Balescu, and Professor P. Resibois. In addition, we thank Dr. Richard Rothery for his continued interest in this work and for many profitable discussions, as well as for the data used in Fig. 1.1 for the speed distribution function. We also thank Dr. Tenny N. Lam for many profitable discussions and suggestions and Mr. Leslie C. Edie for his interest and for kindly providing us with the unpublished Port of New York Authority manuscript by J. L. Kraus. It is indeed a pleasure to acknowledge with thanks the assistance of Miss Margaret Malkowski in typing various versions of the manuscript. Finally, we express our gratitude to the General Motors Research Laboratories for its generous support of this research, and to Dr. L. R. Hafstad, former vice president of the Laboratories, Mr. J. M. Campbell, former scientific director of the Laboratories, and Dr. P. F. Chenea, vice president of the Research Laboratories, for their interest, support, and encouragement.

Introduction

During the past few decades we have witnessed an enormous growth of the branch of physics that deals with the behavior of large numbers of atoms or molecules. This type of physics is generally called "statistical physics." The first form of statistical physics was probably the kinetic theory of gases as formulated in modern times. On the whole, however, classical physics was mainly concerned with systems involving only a few degrees of freedom, as exemplified by the study of planetary motion or the motion of a pendulum. In recent years, the development of methods to treat many-body systems has led not only to a better understanding of the properties of matter, as for example specific heat or viscosity, but also to a clearer understanding of conceptual features, such as the meaning of irreversibility, the organization of macroscopic systems, and the approach to equilibrium or to steady states. We now ask: Is it possible to conceive a statistical physics in which the "objects" would be human beings instead of molecules? Clearly the following remarks have to be kept in mind.

In statistical physics all quantities characterizing single molecules or collections of only a few molecules, such as mass, moments of inertia, and intermolecular forces, enter as *given* parameters. It is in terms of these parameters that quantities that have a meaning only for many-body systems (e.g., the melting properties of a solid or the diffusion coefficient for a liquid mixture) have to be finally expressed. We have a clear-cut distinction between "individual" and "collective" properties. (In fact, this distinction is not so simple, even in physics. The mass of a particle is related to the field the particle generates, and the field is a system with a large, even infinite, number of degrees of freedom.)

Now let us consider one of the simplest problems that involves the behavior of human collectivities and that can be treated mathematically. Suppose that a car arrives at a stop sign at time $t = 0$ and the driver observes cars traveling on a main intersecting road that arrive at the intersection in a random way. The interval between the arrival times generates the gaps between the cars. The waiting driver is delayed a random time until he accepts some gap. We wish to calculate the statistics of delay to the stopped driver. The basic role is played by the gap acceptance function $\alpha(t)$, defined as the probability that the waiting driver will merge if the gap is of size t.

Direct measurements have shown (Herman and Weiss, 1961) that the gap acceptance probability $\alpha(t)$ may be approximated analytically as follows:

$$\begin{aligned} \alpha(t) &= 0, && t \leq T, \\ &= 1-\exp[-\alpha(t-T)], && t > T. \end{aligned}$$

This is a problem that has close analogues in statistical physics. For example, think about the probability of a chemical reaction between molecules belonging to crossing molecular beams where the function $\alpha(t)$ would be an "individual" property. In the present case, however, this is no longer exactly so, as the gap acceptance process depends on rules and patterns of behavior that are characteristic of societies. An "average" Italian driver would perhaps accept a gap that the "average" American driver might still refuse!

The distinction between individual and collective behavior is much less clear than in physics. Whereas in physics molecules of the same species are exactly alike (they have, e.g., the same mass, etc.), even the individual properties of a human population are subject to statistical fluctuation. However, it is fortunately possible to formulate an analogue to conventional statistical physics, but in a slightly different way.

Suppose we know the behavior of a population under given conditions, for example, the behavior of drivers in sufficiently dilute traffic on a highway, so that we may neglect their interactions as we can neglect interactions in the economic behavior of Robinson Crusoes, each isolated on his own island. We wish then to develop a statistical theory that would permit us to predict the behavior of this population under modified conditions, that is, taking into account interactions between drivers or trade between the Robinson Crusoes.

To the individual properties of statistical physics now correspond the patterns of behavior of the population in simplified limiting conditions. The theory of traffic flow described in this monograph is precisely of this type. We wish to predict the pattern of traffic at arbitrary density in terms of the driver behavior in dilute, noninteracting traffic. Even in the framework of this formulation there are some essential differences that remain. Human beings have a purpose, or microgoals as economists often call them (see, e.g., Lowe, 1965), as well as memory. Clearly, the mathematical description used in the "statistical physics of human behavior" has to incorporate these features. The mathematical description of men will always be very different from the mathematics of molecules.

Is it at all possible to incorporate such features without making the theory so cumbersome that it loses all its power of prediction? There is, of course, no general a priori answer. The only reasonable possibility, it seems to us, is to try to develop statistical theories for relatively simple situations, which possibly can then be expanded. In this perspective, it seems especially interesting to consider traffic flow problems.

The main reasons why traffic flow theory may be a good testing ground for the construction of a more general statistical theory of human behavior appears to us to be the following. Traffic flow can be measured, in principle, in the same way as can be measured any flow of fluid; also, the relevant parameters that determine the value of the flow (such as density of traffic, or number of lanes on a highway) can be identified and varied. Moreover, each driver has generally only one single purpose, namely, to perform his journey in a "reasonable" length of time; this contrasts with the multidimensionality of most other aspects of social life. For all these reasons the traffic problem appears as one of the most amenable to mathematical treatment.

At the same time, the traffic problem presents features that invite the formulation of a statistical theory in the sense described in the earlier part of this introduction. Indeed, the basic problem here is the competition between the drivers, the conflict between the goal of a given driver and the goals of the drivers who interact with him through passing or other maneuvers. The competition in this case is directly measurable, as it gives rise to a characteristic reduction in the average speed of the driver population with increasing concentration on the highway. Now what are some types of approach that have been proposed to deal with this problem? (For short review articles see Herman and Gardels, 1963; Weiss, 1965; Herman,

1966; and Gazis, 1967. See also the *Proceedings* of the international symposia on the theory of traffic flow: Herman, 1961; Almond, 1965; Edie *et al.*, 1967; Leutzbach and Baron, 1969; as well as the Australian Road Research Board Proceedings, 1962, 1964, 1966, 1968. Also of interest are Haight, 1963; Ashton, 1966; Road Research Laboratory, 1965; Gerlough and Capelle, 1964; Horton, 1965. The journals *Transportation Science*, *Transportation Research*, and *Operations Research* are also noteworthy.)

Attempts have been made to derive a continuum hydrodynamics of traffic flow somewhat similar to classical hydrodynamics. Certainly one of the most important contributions in that direction is the work of Lighthill and Whitham (1955). We have summarized their approach in Chapter 2.

As in classical hydrodynamics or thermodynamics, results may be reached in this way, but we are often left with a feeling of dissatisfaction. Indeed, by using such a method, no direct connection can be made between the behavior of the population and the characteristics of the traffic flow. It is merely shown that some characteristics of homogeneous time-independent traffic as described by the flow-concentration curve imply well-defined consequences for inhomogeneous situations, namely, "kinematic waves."

At the other extreme are theories based on the individual behavior of vehicles. An especially successful version is the so-called follow-the-leader theory first discussed by Reuschel (1950) and Pipes (1953) but most extensively developed beginning with the work of Chandler *et al.* (1958). We have given a very brief summary of some relevant aspects of car-following theory in Chapter 1. In this theory, however, only a single lane of traffic is taken into account and passing is not allowed. In Miller's work (1961 a, b, 1962, 1963, 1967) traffic is described as a succession of traveling queues with some rules prescribed for faster queues catching up with slower ones and then passing them. It is clear that such theories have to face an enormous mathematical complexity as a consequence of the large variety of situations that exist for a given number of cars under passing conditions. The theory we will describe occupies an intermediate position between hydrodynamic theories of traffic and theories based on the behavior of individual driver–car units.

As already mentioned earlier in this introduction, our approach follows closely the pattern of standard statistical physics. We wish to describe the behavior of the driving population for a given traffic density in terms of the assumed known behavior at sufficiently low concentration. In order to accomplish this we introduce, as is customary in statistical physics, a phase-space description involving both the positions and the velocities of the cars. Our main goal is the derivation and discussion of a kinetic equation that describes the space-time evolution of the velocity distribution of cars (Prigogine, 1961).

The most remarkable feature is probably that in the framework of this type of description we may incorporate readily the most essential differences between the behavior of a population of drivers and that of ordinary molecules! Contrary to the standard kinetic equations for molecules, where there appears only one type of term (homogeneous situations) corresponding to the interaction between the molecules, we may now introduce two types of terms: one corresponding to the interactions and the other to the "desires" of the individual drivers to realize a given goal. Clearly terms of this type have no analogue in molecular physics. It is the combined effect of these two terms that expresses the competition between the will of a given driver and the influence of the other drivers who interact with him.

In our kinetic approach we introduce quantities such as the probability of passing and relaxation times. For purposes of numerical calculation we replace them with functions of the traffic density using simple plausibility arguments regarding their functional form. At the present time it is difficult to imagine a theory from which it would be possible to "derive" such quantities. It seems more plausible that only the accumulation of experimental data could give us the necessary supplementary information. Also, we hope that a comparison between the predictions of our theoretical model and experimental data would throw some light on such questions.

Statistical theory involving human behavior will always present profound differences when compared with statistical physics. There are as yet no laws in this area comparable to the laws of classical or quantum mechanics from which it is possible to deduce the values of the parameters that appear in the theory.

In spite of the lack of experimental and observational data, we believe it would be appropriate to present a survey of our theoretical results. The study of our kinetic equation has led to a certain number of unexpected and, we believe, interesting conclusions.

As shown in detail in Chapter 4, we are led to a clear distinction between two types of flow regimes, which we designate, respectively, the individual and the collective flow regimes. The collective flow regime corresponds to a kind of traffic "condensation" occurring at a critical concentration. There the flow patterns become largely independent of the will of individual drivers; they express directly the overall behavior of the collectivity.

Furthermore, the study of the kinetic equation for inhomogeneous systems (see Chapter 5) has on the one hand clarified the relationship of the statistical theory with hydrodynamics and on the other hand has led to a new theory of traffic stability. We are able to follow in detail why, at increasing traffic density, instabilities must arise and then to study the types of perturbation waves generated by local inhomogeneities. Among the different types of perturbations we can identify the kinematic waves of the hydrodynamic theory mentioned earlier.

All these conclusions are largely independent of the detailed form of the kinetic equation. For this reason we believe that they will remain valid even if future observations were to force us to modify the precise form of the kinetic equation. One of the reasons for this confidence is that the theory appears sufficiently flexible to incorporate nonlinear effects that describe "adaptive" behavior of the driver population.

This is an essential point that every satisfactory statistical theory of human behavior has to incorporate. We remarked at the beginning of this introduction that we suppose the behavior of the population to be known under idealized conditions i.e., for infinite dilution on the highway, and that we wish to use this knowledge, combined with the interactions, to make predictions that are valid under slightly different conditions i.e., for a given finite concentration. But clearly the population may change its behavior, that is, change its "strategy," when the conditions are altered.

Again we find that the distinction between individual and collective properties is much less obvious than in physics. We could really consider driving on a highway as a kind of game in which the number of players is practically infinite. However, there is an essential difference between this type of game and games, such as chess or bridge, where the rules are given once and for all time. In the case of vehicular traffic, the "rules" change according to the circumstances.

We have, therefore, a continuous feedback between "purpose"

and "realization," which is studied in detail in Chapter 6 and which is likely to appear as well in more complicated problems of sociology or economics.

In this monograph we have employed a level of presentation that we hope will not lead to difficulties for most readers. Only in Chapter 5 have we presupposed some specialized knowledge of stability theory based on dispersion equations. That chapter, however, is not essential for the understanding of the general features of our approach.

We have to admit regretfully that at present the existing experimental material does not permit us to properly test our theory in a quantitative way nor to verify the assumed form of the kinetic equation. We hope that because of the inherent interest of the subject a systematic effort will be made in this direction during the coming years.

CHAPTER 1

Basic Concepts

1.1 Flow-Concentration Diagram

Consider a highway on which passing is allowed to occur. We may assume that there exists at a given time and at a given point a velocity (the terms velocity and speed are used interchangeably throughout this monograph) distribution function $f(x, v, t)$. In terms of this function, the number of cars dN that at a given time t

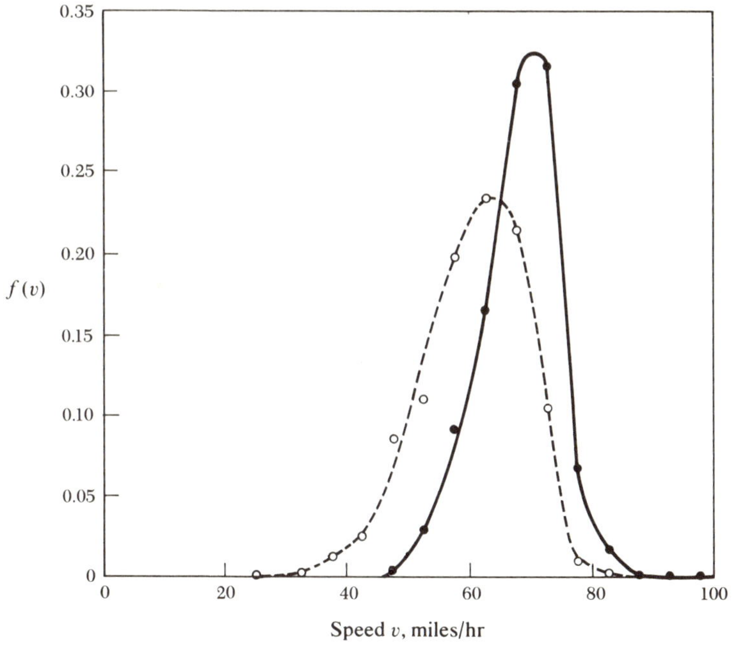

FIG. 1.1. Observed speed distribution function $f(v)$ versus speed. The data were obtained at one position on a 3.5-mile stretch of the Lansing Highway, I-96, on which there were no entrances or exits. The closed circles represent $f(v)$ for the left lane (644-car sample) and the open circles, the right lane (782-car sample).

are in the road interval between x and $x+dx$ and in the velocity interval between v and $v+dv$ is given by

$$dN = f(x, v, t)\, dx\, dv. \tag{1.1}$$

Unfortunately, only a few measurements of the distribution function have been made. Some experimental data are represented in Fig. 1.1.

Once the velocity distribution function is known, we may derive important quantities, such as the local concentration $c(x, t)$,

$$c(x, t) = \int_0^\infty dv\, f(x, v, t) \tag{1.2}$$

and the local flow q,

$$q = c\bar{v}(x, t) = \int_0^\infty dv\, vf(x, v, t). \tag{1.3}$$

We will also consider very often the velocity dispersion, given by

$$c\overline{(v-\bar{v})^2} = \int_0^\infty dv\, (v-\bar{v})^2 f(x, v, t). \tag{1.4}$$

All these quantities are moments of the distribution function f. (According to standard terminology, the moments M_n associated with a probability distribution function $\psi(x)$ are defined as $M_n = \int dx\, x^n \psi(x)$.) Of special importance are, of course, the concentration and flow, which may be measured directly on highways. It may be useful to introduce a graphical representation of the x, v space as shown in Fig. 1.2. In this figure we have represented the infinitesimal element of area $dx\, dv$. It is very important to remember that this element of area must be large enough to contain a sufficient number of cars so that f (and correspondingly c and $c\bar{v}$) may be treated as continuous functions of their arguments.

If fluctuations of concentration arise through formation of queues, the local element dx must be chosen large with respect to the size of the fluctuations. All this is familiar from physics. In order to speak of the concentration of a fluid, we have to take a volume element and then count the number of particles that are located inside. Then to obtain a well-defined result, the volume element has to be large with respect to the size of density fluctuations. At the same time, if the fluid is inhomogeneous, we have to make the size of the volume element small with respect to the parameter characterizing the inhomogeneity.

Going back to formulas (1.2) and (1.3), it is natural to assume that on any uniform stretch of the road the flow $q = c\bar{v}$ is a function

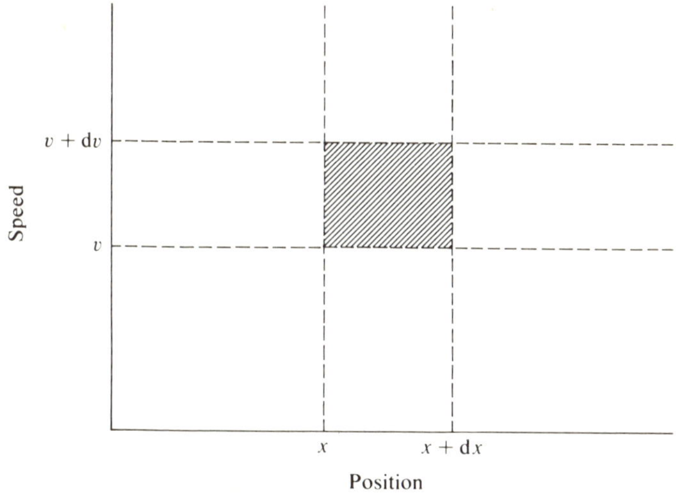

FIG. 1.2. Schematic representation of the x, v space.

of the concentration c. This is in agreement with a large amount of information that is now available. It is the starting point of the hydrodynamic theory of traffic as discussed in Chapter 2. Figures 1.3 through 1.5 illustrate the relationship between flow and concentration and between speed and concentration. In all cases the qualitative form of the flow-concentration curve is the same. For low concentration the flow grows approximately linearly with the concentration; then it begins to grow less rapidly, goes through a maximum, and finally falls to zero in some manner until the "jam" concentration c_p is approached.

An interesting feature of the flow-concentration curve is the occurrence of a large dispersion in the flow for high concentrations ($c > \sim 50$ cars/mile). We are reminded of the well-known distinction between laminar and turbulent flow in classical hydrodynamics. Thus, as we will see later (in Chapter 5), this analogy can indeed be justified by the theory of traffic flow. It will be shown that at high concentrations there exist instabilities that are formally very similar to those responsible for turbulence in classical hydrodynamics.

The flow-concentration curve has often been called the "fundamental diagram" of road traffic (see, e.g., Haight, 1963, Chapter III), whereas other authors term it the "equation of state" of traffic theory. The flow-concentration curve is fundamental in the

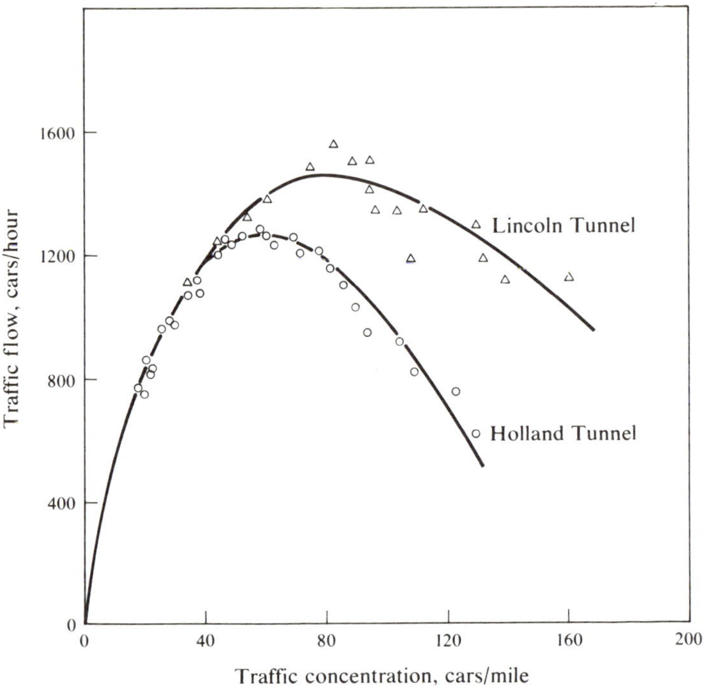

FIG. 1.3. Flow (cars per hour) versus vehicle concentration (cars per mile) for the Lincoln and Holland Tunnels. The Lincoln Tunnel data are those of Greenberg (1959). The solid curves correspond to least squares fits of the reciprocal-spacing car-following model to the data.

sense that it is at least implicit in every flow theory of traffic under a variety of assumptions.

The interpretation of the qualitative behavior of the flow curve is clear. Indeed it can be seen from Fig. 1.4 that the average velocity versus concentration curve should be of the form shown in Fig. 1.5. For low concentrations the average velocity would be a constant equal to the "mean free speed," since for dilute traffic the drivers behave independently. For higher concentrations, however, the average velocity begins to decrease with increasing concentration as a result of the mutual hindering of the drivers. This means, of course, that the entire distribution function $f(x, v, t)$ has been modified by the interaction between the drivers. An actual speed versus concentration curve is shown in Fig. 1.6. Clearly the theoretical models

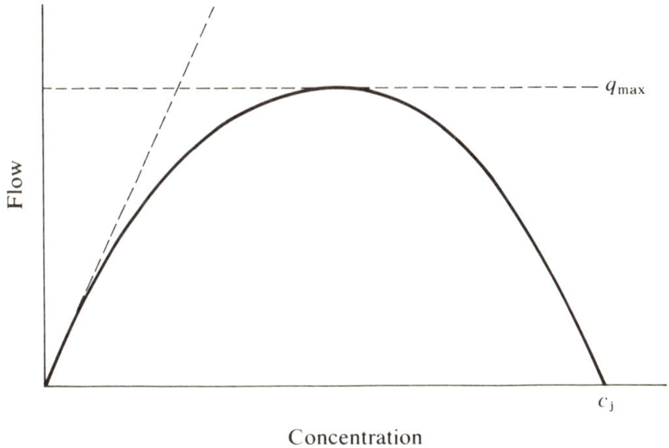

FIG. 1.4. Schematic diagram of flow versus concentration.

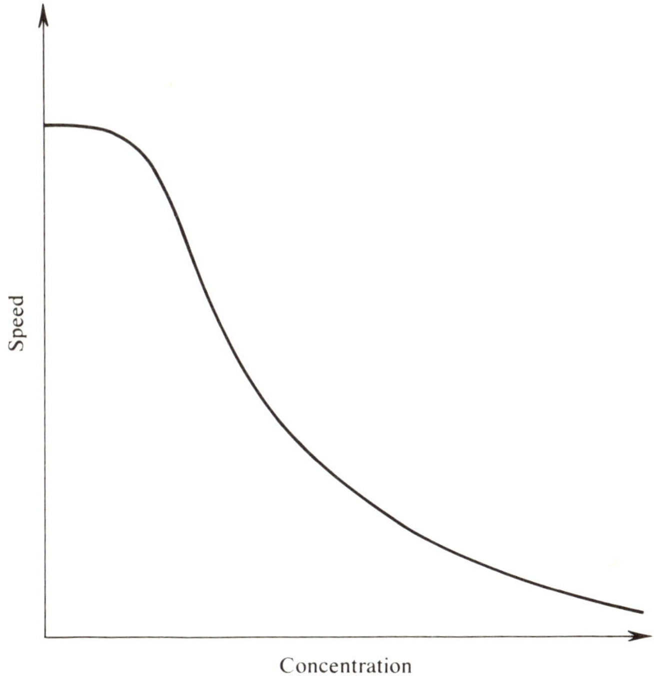

FIG. 1.5. Schematic representation of the speed versus concentration curve.

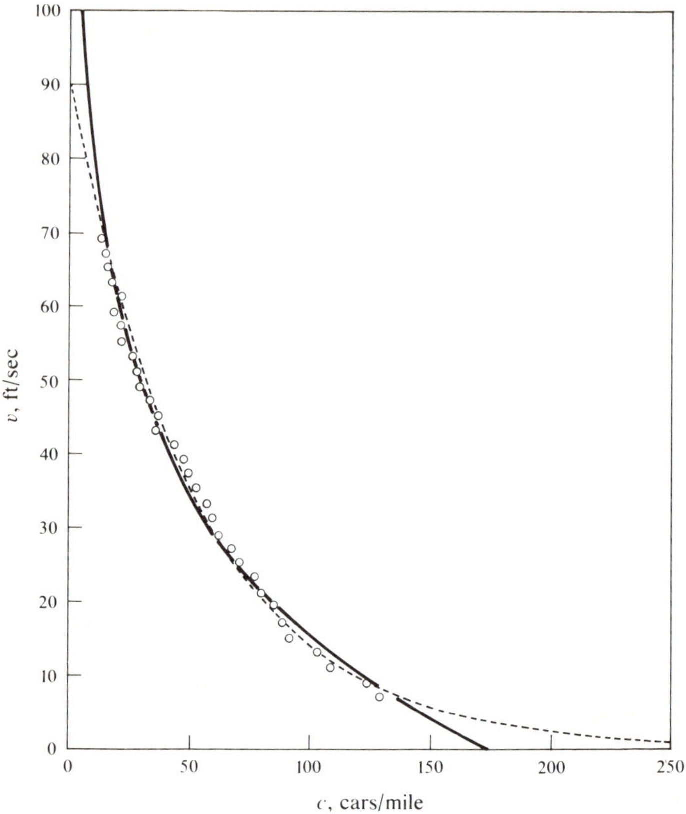

FIG. 1.6. Data on speed (feet per second) versus vehicle concentration (cars per mile) plotted together with two speed–concentration relations derived from car-following models. The data were obtained in the Holland Tunnel. The solid curve is based on the reciprocal-spacing car-following model and the dotted curve is from Edie's model (see Section 1.2).

do not fit the observed data in the region of very low concentrations, where observations indicate that the speed of the traffic stream is independent of vehicle concentration. If a sufficient number of observations are made in the region $k \simeq 0$, we find that $u = u_f$, the mean free speed.

Here we are already faced with the basic problem associated with every theory of traffic. Can we describe quantitatively the so-called friction that results from the competition between drivers? In other words, what determines the specific form of the flow curve? Also, once the flow curve is established for a homogeneous highway, may we use it to describe inhomogeneous situations such as may arise, for example, when a traffic concentration perturbation is propagated?

1.2 Follow-the-Leader Model and the Flow-Concentration Diagram

The follow-the-leader theory attempts to describe the behavior of a single lane of fairly dense traffic in terms of the detailed manner in which driver units follow one another in a traffic stream, with passing not allowed. This problem has been analyzed in terms of the response of humans to certain external stimuli. The basic assumption of the car-following theory of traffic flow along a single lane of highway is that each vehicle in the line of traffic follows the one in front of it according to a stimulus–response relationship; that is, response at time $t+T$ equals a sensitivity coefficient or gain factor multiplied by the stimulus at time t. The various mathematical formulations of the ideas that have been given can be stated in the following form:

$$\frac{d^2x_n(t+T)}{dt^2} = \lambda\left[\frac{d^px_{n-1}(t)}{dt^p} - \frac{d^px_n}{dt^p}\right]. \tag{1.5}$$

It seems most reasonable, and it is borne out by experiments, that stimuli involving relative acceleration (or jerk, etc.), that is $p > 1$, are of little importance since drivers cannot make accurate judgments on these quantities. It has been shown theoretically that following by means of relative spacing alone is unstable (Herman *et al.*, 1959; see also R. L. Bierley, 1963), which rules out the case $p = 0$, a point that has been verified by experiment. It turns out that the experimental data can be quite well represented with an equation of the form (1.5) with $p = 1$,

$$\frac{d^2x_n(t+T)}{dt^2} = \lambda\left[\frac{dx_{n-1}(t)}{dt} - \frac{dx_n(t)}{dt}\right] \tag{1.6}$$

with the sensitivity coefficient λ in general given by (see Gazis *et al.*, 1961)

$$\lambda = \frac{\lambda_0[dx_n(t)/dt]^m}{[x_{n-1}(t) - x_n(t)]^l}. \tag{1.7}$$

Two major classes of phenomena are of special interest, namely, stability and steady-state flow. We will discuss here only the problem of the steady-state flow, which leads to a simple analytic form of the flow-concentration curve. On an empty highway and on a highway jammed with bumper-to-bumper traffic it is clear that there is zero flow. If there exists a unique flow versus concentration relationship, then there is a concentration between zero and jam concentration for which the flow is a maximum.

Let us assume that a platoon of vehicles undergoes a transition from one steady-state situation to another after a slowing-down or speeding-up maneuver of the lead car. Then we can integrate the car-following equation from zero to infinity, which disposes of the time lag. In the simplest possible case, for which $l = m = 0$ in Eq. (1.7) or λ = constant = λ_0, we have the linear car-following situation. After integration of Eq. (1.6) we obtain

$$\frac{dx_n(t)}{dt} = \lambda_0[x_{n-1}(t) - x_n(t)] \tag{1.8}$$

or

$$v = \lambda_0 s + \beta \tag{1.9}$$

where s is the spacing $x_{n-1} - x_n$ and β is a constant determined by the boundary conditions when $v = 0$, $s = s_j$. The quantity j refers to the jammed state, so that

$$v = \lambda_0(s - s_j) = \lambda_0\left(\frac{1}{c} - \frac{1}{c_j}\right) \tag{1.10}$$

in which the concentration $c = s^{-1}$.

The transition from the initial follow-the-leader equation including the time lag to the flow-concentration equation involves a supplementary time average. This implies that the situation described is, on the average, time independent (stationary process). Since the steady-state flow is given by $q = vc$, we have

$$q = \lambda_0(1 - c/c_j). \tag{1.11}$$

It is obvious from the assumptions made that the follow-the-leader theory cannot be applied to the dilute state. Therefore it is not surprising that Eq. (1.11) for the flow does not agree with the experimental finding that the flow-concentration curve should pass through the origin and approximately follow a linear relationship in the vicinity of the origin.

Another interesting choice of the sensitivity coefficient corresponds to the nonlinear car-following case for which $m = 0$ and $l = 1$. If we follow the procedure above (Gazis *et al.*, 1959) and integrate, we obtain for this case the result

$$v = \lambda_0 \, in \, s + \text{constant}, \tag{1.12}$$

which becomes, after noting that $v(c_j) = 0$,

$$v = \lambda_0 \, ln(s/s_j) = \lambda_0 \, ln(c_j/c) \tag{1.13}$$

with

$$q = cv = \lambda_0 c \, ln(c_j/c). \tag{1.14}$$

A considerable amount of data has been obtained on flow versus concentration for the single-lane case, especially in the vehicular tunnels in New York City (see Brainard *et al.*, 1961; Edie *et al.*, 1963). The flow data on a sample of over 20,000 cars in the Holland Tunnel in New York City fit the foregoing nonlinear flow equation rather well. Good fits have similarly been obtained with data from other vehicular tunnels (see Edie *et al.*, 1963). The speed at maximum flow, called the characteristic speed, is interestingly the parameter λ_0 in Eq. (1.13). A comparison of the characteristic speed obtained in various vehicular tunnels by means of microscopic experiments on car following with that obtained from the flow curves show striking agreement. In the Lincoln Tunnel we find $\lambda_0 = 20$ mph, compared to $v_{\max} = 17$ mph; Holland Tunnel $\lambda_0 = 18$ mph, $v_{\max} = 19$ mph; Queens Midtown Tunnel $\lambda_0 = 22$ mph, while $v_{\max} = 25$ mph. It is interesting to note that the characteristic speed is higher in the newer tunnels where greater lane width, improved lighting, and so on, may play a significant role in allowing the driver to use a shorter spatial headway for a given speed.

Various attempts have been made to improve the fit of the theory to the flow data, for example, by Edie (1961) and Gazis *et al.* (1961). We have examined the more general sensitivity coefficient indicated earlier and found in view of the available data that it was not possible to establish definite superiority of any one particular model. The nonlinear model, however, is required to account for the flow versus concentration data, but it is not clear that a somewhat more complicated nonlinear model has any distinct advantages over the simple nonlinear one; the basic features and interpretations of the flow curve are not altered.

Clearly, all the car-following equations represent averages taken over the population of drivers. Therefore all the quantities, such as the speed and flow, must be understood as ensemble averages. (The ideas in this section are summarized in a review by Herman, 1966.)

CHAPTER 2

Continuum Hydrodynamic Theory

2.1 Kinematic Waves

A very interesting macroscopic hydrodynamic description of traffic flow has been developed by Lighthill and Whitham (1955; see also Richards, 1956). We present a summary of their work, as well as a short discussion of traffic hydrodynamics, before we proceed in the next chapter with the development of the statistical theory of traffic.

The main result of the Lighthill–Whitham theory is the derivation of kinematic waves that satisfy a first-order partial differential equation. The terminology "kinematic waves" is used here, in the sense of Lighthill and Whitham, to mean that the existence of these waves is not derived from a detailed microscopic dynamic law, such as a Newtonian law in classical mechanics, but follows directly from the flow-concentration relationship described in Chapter 1.

We start with the continuity equation

$$\frac{\partial c}{\partial t} + \frac{\partial c\bar{v}}{\partial x} = 0, \tag{2.1}$$

which expresses simply that there is no creation or destruction of cars on the road; the partial derivative with respect to time of the concentration is given by a divergence of the flow.

A similar equation is valid in ordinary fluid hydrodynamics and kinetic gas theory (see, e.g., Landau and Lifshitz, 1959). However, traffic hydrodynamics departs radically from ordinary fluid hydrodynamics in that it is now assumed that there exists a functional relation between flow and concentration, namely,

$$c\bar{v} = q\,(c). \tag{2.2}$$

We have seen that the existence of such a relationship is in agreement with experimental data for homogeneous situations (see Section 1.1). In this chapter we assume that this functional dependence is valid even for space and time-dependent situations. As a consequence, Eq. (2.1) may be written

$$\frac{\partial c}{\partial t} + V\frac{\partial c}{\partial x} = 0 \tag{2.3}$$

with

$$V = \frac{\partial(c\bar{v})}{\partial c} \,. \tag{2.4}$$

In this manner, we obtain a first-order partial differential equation that has solutions of the form

$$c = c(x - Vt). \tag{2.5}$$

As a consequence, an inhomogeneity, for example, a traffic "wave," will be propagated with a velocity V that is obviously different from the average velocity $\bar{v}$. We will call phenomena of this type, which propagate with velocity V, *kinematic waves*, and the quantity V the *kinematic wave velocity.*

As a result of the form of the flow-concentration curve, the kinematic wave velocity V is always smaller than $\bar{v}$. It is given by

$$V = \bar{v} + c\frac{\partial \bar{v}}{\partial c} \leq \bar{v} \tag{2.6}$$

which follows from the fact that $\partial\bar{v}/\partial c \leq 0$. In the limit of small concentrations, $V \to \bar{v}$; for concentrations corresponding to the flow curve maximum, $V = 0$; and for concentrations that are higher, $V < 0$. As the kinematic wave velocity decreases with concentration, successive waves may catch up and form "discontinuity waves." By analogy with fluid dynamics, the term shock waves is often used in the traffic literature.

As is well known (see, e.g., Hirschfelder *et al.*, 1964), the very existence of acoustic waves in continuum physics is based on the maintenance of the Maxwell velocity distribution. Indeed, if we apply a periodic disturbance to a gas, the collisions between the molecules restore the Maxwell distribution in a characteristic time τ_{rel} generally called the relaxation time. Under usual conditions the relaxation time is much smaller than the characteristic time associated with the perturbation of the gas.

We will see in Chapter 5 that the statistical theory of traffic flow leads to a similar interpretation of kinematic waves. As in the case of acoustical waves, they express the coherence of the medium in response to perturbations. The mathematical theory of kinematic waves is much simpler than the ordinary theory of waves based on continuum dynamics. For example, the average velocity, which is an independent variable in ordinary continuous mechanics, now becomes, through Eq. (2.2), a function of the concentration. Therefore, we may write for the local average velocity $\bar{v}(x, t)$

$$\bar{v}(x, t) = \bar{v}[c(x, t)]. \tag{2.7}$$

Let us indicate a few consequences of this equation. We now have a well-defined relationship between the average acceleration of an element of the traffic flow and the concentration gradient. Indeed, from Eq. (2.7) the average acceleration for an observer moving with the average speed of the traffic fluid can be written

$$\begin{aligned} \frac{d\bar{v}}{dt} &= \frac{\partial \bar{v}}{\partial t} + \frac{\partial \bar{v}}{\partial x}\bar{v} \\ &= \frac{\partial \bar{v}}{\partial c}\frac{\partial c}{\partial t} + \frac{\partial \bar{v}}{\partial c}\frac{\partial c}{\partial x}\bar{v}. \end{aligned} \tag{2.8}$$

Combining the conservation of vehicles as expressed by Eq. (2.1) with Eq. (2.8) yields for the average acceleration

$$\frac{d\bar{v}}{dt} = \Phi \frac{\partial c}{\partial x} \tag{2.9}$$

where the friction coefficient Φ is given by

$$\Phi(c) = -c\left(\frac{\partial \bar{v}}{\partial c}\right)^2 \leq 0. \tag{2.10}$$

Equation (2.9) gives the acceleration of the traffic fluid in terms of the concentration gradient and can serve as a convenient starting point for the comparison between the dynamics of the traffic fluid and that of a classical fluid.

Since the friction coefficient $\Phi(c)$ is a negative quantity, the average acceleration for an observer moving with the average speed of the traffic fluid is positive when moving into a less dense region of traffic and negative when moving into a denser region of traffic. It is of interest to note that some authors have postulated relationships of the form of Eq. (2.9) (see, e.g., Greenberg, 1959). This is not necessary, as it is evident that this type of equation is a direct consequence of number conservation and the steady-state assumption.

We now introduce a second friction coefficient, ϕ, which relates the average acceleration for an observer stationed at a fixed position on the roadway to the concentration gradient. Indeed from Eq. (2.7), we have

$$\frac{\partial \bar{v}}{\partial t} = \frac{\partial \bar{v}}{\partial c} \frac{\partial c}{\partial t} \tag{2.11}$$

which we can combine with Eq. (2.3) to obtain

$$\frac{\partial \bar{v}}{\partial t} = \phi \frac{\partial c}{\partial x} \tag{2.12}$$

where

$$\begin{aligned} \phi &= -\frac{\partial \bar{v}}{\partial c} \frac{\partial (c\bar{v})}{\partial c} \\ &= -\frac{\partial \bar{v}}{\partial c}\left(\bar{v} + c\frac{\partial \bar{v}}{\partial c}\right) = \Phi + \bar{v}\left(-\frac{\partial \bar{v}}{\partial c}\right). \end{aligned} \tag{2.13}$$

A schematic representation of the friction coefficient ϕ versus concentration is given in Fig. 2.1. It should be noted that ϕ is zero at

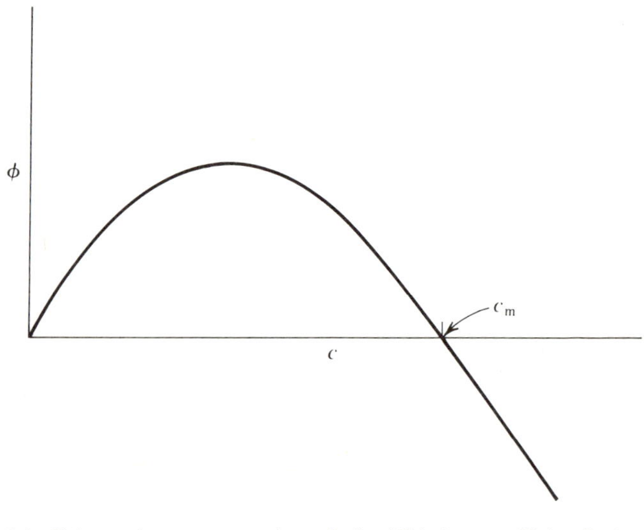

FIG. 2.1. Schematic representation of the "friction coefficient" ϕ versus concentration.

the origin, since for low concentration $\partial v/\partial c \sim 0$. This follows because we would expect that $\bar{v}$ would start out at some value $\bar{v}^0$ plus a term of order at least c^2 because of binary interactions between drivers. Then ϕ increases to a maximum and vanishes again at c_m, which is the concentration at which the flow-concentration curve is a maximum. Hence, for an observer at a fixed point on the roadway with $c < c_m$, we have corresponding to a positive concentration gradient an increase of the average speed. This is an unavoidable consequence of the steady-state assumption (see Eq. 2.7), which shows that in this region $\partial c/\partial t$ and $\partial c/\partial x$ are of opposite sign.

It may be of interest to relate the macroscopic speed obtained from the nonlinear model of the car-following traffic theory (Gazis *et al.*, 1959) to the present discussion. For this model (see Section 1.2)

$$\bar{v} = -\lambda \ln c + \text{constant} \tag{2.14}$$

where λ is the speed at maximum flow. In this case we find that the friction coefficients are given by

$$\Phi = -\frac{\lambda^2}{c} \tag{2.15}$$

and

$$\phi = \frac{\lambda}{c}(\bar{v}-\lambda). \tag{2.16}$$

We see from a comparison of Eq. (2.16) with the general required behavior of ϕ that Eq. (2.16), as previously noted, does not have the correct limiting behavior as $c \to 0$.

The discussion above serves to illustrate further the difference in structure between the macroscopic hydrodynamics of fluids and of traffic flow. In the case of fluids we have to supplement the information given by thermodynamics (equation of state, etc.) with five supplementary independent equations (for the simplest case of reversible motion). In traffic flow we have to supplement the information given by the study of homogeneous time-independent situations with only *one* supplementary equation, which is the equation of continuity (2.1). We will discuss this matter in greater detail in Chapter 3.

2.2 Discussion of the Hydrodynamic Approach

As we have just indicated (in Section 2.1), the simplicity of the theory of traffic from the point of view of hydrodynamics originates from the fact that it is based solely on the conservation of the number of cars. Other conservation laws, such as conservation of energy or momentum, have no analogue in traffic hydrodynamics. This is an essential distinction which radically differentiates both the hydrodynamic and the statistical theories of traffic flow from the corresponding disciplines in "ordinary" physics.

In spite of its interest, the hydrodynamic approach to traffic problems can hardly be considered the final formulation of traffic theory. First, it gives no indication whatsoever about the form of the flow-concentration relation (2.2). Also, it assumes that this relationship may be used even under time- and space-dependent conditions. However, there are conditions that must be satisfied in order to apply (2.2). These conditions refer to the size of the inhomogeneities and to the time scale involved. Thus we will see in detail in Chapter 5 that when the size of the inhomogeneity becomes too small (i.e., the concentration gradient becomes too abrupt), the relationship (2.2) is no longer valid. Also, the hydrodynamic theory of traffic neglects the deformation of kinematic waves. Again, as we will see, this is only permitted for times that are not too long. Finally, the hydrodynamic approach neglects completely the role of instabilities, which, specifically at higher concentrations, may become dominant. Extensions of the hydrodynamic theory have been proposed. For example, Eq. (2.2) can be generalized (Franklin, 1965) to include space and time derivatives of the concentrations,

$$c\bar{v} = q = F\left(c, \frac{\partial c}{\partial t}, \frac{\partial c}{\partial x}\right). \tag{2.17}$$

However, the physical meaning of such an extension is not clear. In the case in which the Lighthill–Whitham theory is inadequate, is it at all possible to express the flow in terms of other average quantities? The natural procedure, as in ordinary fluid dynamics, seems to be to turn to a statistical theory for deeper insight into the meaning of continuum theory, which is what we will attempt to do next.

CHAPTER 3

Kinetic Equation

3.1 Low-Density Traffic

The main purpose of this chapter is to obtain an equation for the velocity distribution function $f(x, v, t)$ introduced in Chapter 1. By analogy with ordinary statistical physics, we will call it the *kinetic equation.* We first consider the statistical properties of low-density traffic on an infinitely long highway (see Weiss and Herman, 1962). It may then be assumed that each car travels at some constant speed; moreover, passing is always possible and occurs without change of speed. Therefore, in this case, we have no interaction between drivers.

Since each driver retains his desired speed in this model, the only modification that will occur in the general distribution function for the cars is a change in the spatial distribution. We will show that whatever the value of the initial concentration $c(x, t = 0)$, we have for sufficiently long times

$$c(x, t) = \text{constant}, \qquad t \to \infty. \tag{3.1}$$

In other words, as time goes on, homogenization of the highway traffic occurs as the consequence of the free passing of the cars. This result may be described in many ways (see, e.g., Weiss and Herman, 1962). As an introduction to the more complex cases studied later, we start with the velocity distribution function $f(x, v, t)$. It is immediately obvious that in the absence of any interaction between the drivers it satisfies the equation

$$\frac{\partial f}{\partial t} + v \frac{\partial f}{\partial x} = 0, \tag{3.2}$$

which expresses that the change in the number of cars in the "surface element" $dx\, dv$ is equal to the difference between the number of cars that enter the surface element and those that leave. The general solution of this equation is

$$f(x, v, t) = f(x - vt, v, t = 0). \tag{3.3}$$

This may be verified by straightforward differentiation. The velocity distribution at time t and at point x is equal to that at some initial time $t = 0$ and at position $x - vt$. During the time interval t we have a simple shift of the distribution function from point $x - vt$ to point x.

Let us represent the initial distribution function $f(x, v, 0)$ by a Fourier integral:

$$f(x, v, 0) = \frac{1}{2\pi}\int_{-\infty}^{+\infty} dk\, e^{ikx}\varrho_k(v). \tag{3.4}$$

Then because of (3.3)

$$f(x, v, t) = \frac{1}{2\pi}\int_{-\infty}^{+\infty} dk \exp[ik(x - vt)]\varrho_k(v). \tag{3.5}$$

For long times the exponential will oscillate more and more violently. Therefore in the limit as $t \to \infty$ the only contribution to this integral will come from the region $k \to 0$. This is the so-called Riemann–Lebesgue theorem (see, e.g., Titmarsh, 1932). Thus we have

$$f(x, v, t) \to \varrho_0(v), \qquad t \to \infty. \tag{3.6}$$

As a consequence, we obtain (3.1) by integration over the velocity. Consider an example (Borel, 1925; Brout and Prigogine, 1956). A distribution function that for $t = 0$ reduces to a delta function in coordinate space and corresponds to a gaussian distribution of velocities may be represented by the formula

$$\begin{aligned} f(x, v, 0) &= \frac{1}{(2\pi\sigma^2)^{\frac{1}{2}}}\,\delta(x) \exp\left(-\frac{v^2}{\sigma^2}\right) \\ &= \frac{1}{(2\pi\sigma^2)^{\frac{1}{2}}}\frac{1}{2\pi}\int_{-\infty}^{+\infty} dk \exp(ikx) \exp\left(-\frac{v^2}{\sigma^2}\right). \end{aligned} \tag{3.7}$$

The parameter σ measures the dispersion of the velocity distribution. We then have, at time t, using (3.3),

$$\begin{aligned} f(x, v, t) &= \frac{1}{(2\pi\sigma^2)^{\frac{1}{2}}}\,\delta(x - vt) \exp\left(-\frac{v^2}{\sigma^2}\right) \\ &= \frac{1}{(2\pi\sigma^2)^{\frac{1}{2}}}\frac{1}{2\pi}\int_{-\infty}^{+\infty} dk \exp[ik(x - vt)] \exp\left(-\frac{v^2}{\sigma^2}\right). \end{aligned} \tag{3.8}$$

We may now follow in detail how the car distribution in space, which was concentrated at $x = 0$ for $t = 0$, becomes more and more uniform. Indeed, if we integrate (3.8) over v, we obtain

$$c(x, t) = \frac{1}{(2\pi\sigma^2 t^2)^{\frac{1}{2}}} \exp\left(-\frac{x^2}{\sigma^2 t^2}\right). \tag{3.9}$$

A characteristic feature in this case is that there exists nothing like a relaxation time independent of the initial concentration. The time necessary to achieve uniformity depends on the dispersion of the velocity. In other words, the homogenization process in space depends on the behavior of the complete distribution function $f(x, v, t)$.

There are many interesting consequences that follow from (3.1). For example, it may be shown that an arbitrary headway distribution must relax to a negative exponential in the limit as $t \to \infty$. Also, the probability of passing n cars in time t is described by a Poisson distribution. These conclusions are indeed in agreement with experimental observations for dilute traffic (Leutzbach and Egert, 1965). We will not develop more details here but rather refer the interested reader to the existing literature (Weiss and Herman, 1962; Haight, 1963).

3.2 Desired Speed Distribution and Relaxation Time

In addition to the velocity distribution $f(x, v, t)$ introduced in Section 1.1, we now introduce a second velocity distribution function $f_0(x, v, t)$, which we call the *desired speed distribution* (Prigogine, 1961). The quantity $f_0(x, v, t)\, dx\, dv$ is therefore the number of cars whose drivers have, at time t and on the road interval dx, derived speeds that lie between v and $v+dv$.

The introduction of this function is a very distinctive feature of a statistical theory of traffic as opposed to statistical physics. Clearly the desired speed distribution is a mathematical idealization of the goals that the driver population strives to realize. It corresponds to the "program" each driver wishes to follow. This program may be perturbed as a consequence of various factors, such as obstacles, weather, or interactions with other drivers. As a result, the actual distribution f will become different from the desired distribution f_0.

Two cases are possible: either the perturbation ceases after some time, or it continues to act for arbitrary times. The first case would correspond, for example, to strong winds acting between $t = -T$ and $t = 0$, while the second case might correspond to a continuously congested highway. We now consider the first case, for which we expect that for $t>0$ the velocity distribution would relax back to f_0. We ask, what mathematical form will this relaxation take? There can be no question at present of any a priori "derivation" of this relaxation law since this would require the existence of a mathematical theory of human behavior far beyond our present capabilities. An experimental approach would be very helpful, but unfortunately we have found no adequate data in the literature.

One of the simplest relaxation laws is expressed by the equation

$$f(v, t) = f_0(v) + [f(v, 0) - f_0(v)] \exp\left(\frac{-t}{T}\right). \tag{3.10}$$

In this case we consider the relaxation to a space- and time-independent desired speed distribution $f_0(v)$. This type of relaxation often appears in statistical physics. One of its main attractive features is that it contains only a single relaxation time T. It is clear that it represents a simplification of the real situation. We could, for example, argue that faster cars would perhaps also relax more rapidly to their desired speed. In this case a velocity-independent relaxation time T would not exist. The relaxation law (3.10) satisfies the differential equation

$$\frac{\partial f}{\partial t} = -\frac{f-f_0}{T}. \tag{3.11}$$

More complicated relaxation laws could be introduced easily. We could, for example, consider either a distribution of relaxation times instead of the single relaxation time T, or nonexponential relaxation, which could lead to more complicated differential or integro-differential equations. This would in turn lead to additional complications in our subsequent developments. (In addition to relaxation times, we could also introduce delay or lag times of the type that appear in the follow-the-leader theory; see Section 1.2.)

If the study of detailed examples were to indicate the need, such refinements as discussed above could be introduced into our theory. At present, however, we do not feel compelled to do so. We therefore continue to use the simple relaxation law (3.10). The reader may, if he is so inclined, work out the theory using other forms of the relaxation law.

Some remarks about the normalization of the desired speed

distribution function are needed. Consider an "inhomogeneous highway," which means that the concentration of vehicles along the highway varies and, in particular, consider two stretches of the highway, one around x_1 and the other around x_2. If the "program" of each driver as expressed by the desired speed distribution function remains independent of the local concentration, we may write

$$f_0(x, v, t) = c(x, t)\tilde{f}_0(v) \tag{3.12}$$

where $\tilde{f}_0(v)$ is a desired velocity probability distribution. We may therefore normalize $\tilde{f}_0(v)$ to unity as follows:

$$\int dv\, \tilde{f}_0(v) = 1. \tag{3.13}$$

For the two stretches of road around x_1 and x_2 we then have

$$dN_1 = f_0(x_1, v, t)\, dx\, dv = c(x_1, t)\tilde{f}_0(v)\, dx\, dv$$
$$dN_2 = f_0(x_2, v, t)\, dx\, dv = c(x_2, t)\tilde{f}_0(v)\, dx\, dv$$

where N_1 and N_2 are the numbers of cars with desired speed v at time t in the regions around x_1 and x_2, respectively. The ratio of the number of drivers with desired speed v is then

$$\frac{dN_1}{dN_2} = \frac{c(x_1, t)}{c(x_2, t)}. \tag{3.14}$$

This ratio is independent of $\tilde{f}_0(v)$ and depends only on the local concentration.

It is evident that our model is greatly simplified when (3.12) can be used. We will base most of the work described in Chapters 4 and 5 on this assumption. However, we emphasize that (3.12) can be only an approximation. If the concentration varies very much, we must expect that the drivers will make adjustments to their actual conditions of driving. In Section 3.6, as well as in Chapter 4, we will describe some ways of taking into account these effects, which introduce an additional nonlinearity into our problem. Here again we deal with a problem in "mathematical behavior theory." What is the adaptability of a population of drivers? Do they try to adhere rigidly to the program they would have followed if they had been driving under ideal conditions, such as low concentration? Or, on the contrary, do they adapt rigidly to the new conditions?

To give a priori answers is, of course, out of the question. We believe that one of the most interesting aspects of our theory is, on the contrary, the possibility of developing in a quantitative form the consequences of various assumptions. The comparison with experimental data may then decide which behavior is, in fact, followed by the population.

3.3 Time Scales

It is clear that the space homogenization process considered in Section 3.1 and the velocity relaxation studied in Section 3.2 correspond, in general, to quite different time scales. The homogenization process may be a long-time phenomenon, whereas the velocity relaxation process would, for low-density traffic, normally involve only a few seconds.

As an illustration, consider the velocity relaxation on an inhomogeneous highway. The distribution function $f(x, v, t)$ now changes both because of the flow term, the term proportional to $\partial f/\partial x$ in (3.2), and the relaxation term (3.11), and we can write

$$\frac{\partial f}{\partial t}+v\frac{\partial f}{\partial x} = -\frac{f-f_0}{T}. \tag{3.15}$$

No interaction between the drivers is taken into account in (3.15). This equation therefore refers to a highway with dilute traffic. It may be expected that after a time of order T we will have $f \approx f_0$. Using (3.12) we see that this does not imply that the concentration on the highway has become uniform. It may still be an arbitrary function of space and time. The approach to the final steady state proceeds, therefore, in the following manner. (a) After a time $t \geq T$, the distribution function will be given by f_0; (b) after a time that is in general much longer, the concentration will become uniform if the conditions (boundary conditions, traffic at entrances, etc.) permit. We then have

$$f(x, v, t) = c\tilde{f}_0(v) \tag{3.16}$$

where c is the average concentration on the highway. This question is discussed in detail in the Appendix to this chapter. However, the limit (3.16) can surely only be realized in dilute traffic when interference between drivers is neglected. Therefore, we now turn to the basic problem of how to describe this interference and the corresponding change in the long-time velocity distribution function.

3.4 Kinetic Equation and Interactions

We now try to establish a kinetic equation sufficiently general to take into account the interactions between the drivers. Clearly Eq. (3.15) has to be supplemented by a term that expresses this interaction, so that we write

$$\frac{df}{dt}=\frac{\partial f}{\partial t}+v\frac{\partial f}{\partial x}=\left(\frac{\partial f}{\partial t}\right)_{\text{rel}}+\left(\frac{\partial f}{\partial t}\right)_{\text{int}}. \tag{3.17}$$

As in the previous sections, the first term on the right-hand side arises from the fact that $f(x, v, t)$ is not equal to $f_0(v)$, the desired velocity distribution. The second term takes into account traffic interactions.

As already mentioned, there are at present neither theoretical arguments nor experimental results that lead to an unambiguous choice for the two contributions to df/dt. Fortunately, the detailed structure of these terms appears to be relatively unimportant as long as some general conditions, which we will discuss in Section 3.5, are satisfied. For the relaxation term we again take an expression of the form (3.11):

$$\left(\frac{\partial f}{\partial t}\right)_{\text{rel}}=-\frac{f-f_0}{T}$$

Now, however, we have to expect that T will be a complicated function of concentration related to the probability of passing (see Section 4.1).

We next consider the interaction term. In order to obtain its form we follow closely the arguments usually made in the kinetic theory of gases (see, e.g., Hirschfelder *et al.*, 1964). We may write (Prigogine and Andrews, 1960)

$$\left(\frac{\partial f_i}{\partial t}\right)_{\text{int}}=\sum_j\left(\Gamma_{ij}^{(+)}-\Gamma_{ij}^{(-)}\right)(1-P), \tag{3.18}$$

in which P is the probability of passing. We have temporarily distinguished between a type of car i, which we are following, and the rest of the cars j over which we sum. The quantity $\Gamma_{ij}^{(-)}\,dx\,dv\,dt$ equals the number of cars subtracted from $dx\,dv$ by interaction with cars of type j in time dt, and $\Gamma_{ij}^{(+)}\,dx\,dv\,dt$ equals the number of cars added to $dx\,dv$ by interaction with cars of type j in time dt. Naturally enough, we have multiplied this interaction term by $(1-P)$, the probability of being unable to pass.

Consider now a car of type i at x with velocity v_i. We wish to calculate the probability that, during dt, car i will interact with a car of type j with velocity $v_j < v_i$. Only such an interaction will contribute to $\Gamma_{ij}^{(-)}$. If we fix our attention on car i, then car j approaches with relative velocity $(v_i - v_j) > 0$. The flow of cars of type j as observed from a car of type i at x is $f_j(x, v_j, t)\,(v_i - v_j)$, and the probable number of cars of type j in $dt\,dv_j$ is integrating over all $(v_i - v_j) > 0$,

$$\int_0^{v_i} dt\, dv_j\, f_j(x, v_j, t)\,(v_i - v_j). \tag{3.19}$$

These must interact with the car i at x with probability

$$f_i(x, v, t)\, dx\, dv_i. \tag{3.19a}$$

Thus

$$\Gamma_{ij}^{(-)}\, dx\, dv_i\, dt = dx\, dv_i\, dt\, f_i(x, v_i, t) \times \int_0^{v_i} dv_j\, f_j(x, v_j, t)\,(v_i - v_j). \tag{3.20}$$

In order to obtain $\Gamma_{ij}^{(+)}$, we again follow car i with velocity v_i at x. We want the probability that this car will be interacted with during dt by a car of type j with velocity $v_j > v_i$. Car j approaches with relative velocity $(v_j - v_i) > 0$. The flow of cars of type j as observed from a car of type i at x is $f_j(x, v_j, t)(v_j - v_i)$, and the probable number of cars of type j in $dt\,dv_j$ is integrating over all $(v_j - v_i) > 0$,

$$\int_{v_i}^{\infty} dt\, dv_j\, f_j(x, v_j, t)(v_j - v_i). \tag{3.21}$$

These must interact with the car i at x with probability

$$f_i(x, v_i, t)\, dx\, dv_i\,. \tag{3.21a}$$

Thus

$$\Gamma_{ij}^{(+)}\, dx\, dv_i\, dt = dx\, dv_i\, dt\, f_i(x, v_i, t) \times \int_{v_i}^{\infty} dv_j\, f_j(x, v_j, t)(v_j - v_i). \tag{3.22}$$

Equations (3.19a) and (3.21a) yield

$$\begin{aligned}\Gamma_{ij}^{(+)} - \Gamma_{ij}^{(-)} &= f_i \int_{v_i}^{\infty} dv_j\, f_j(v_j - v_i) - f_i \int_0^{v_i} dv_j\, f_j(v_i - v_j)\\ &= f_i \int_0^{\infty} dv_j\, f_j v_j - f_i v_i \int_0^{\infty} dv_j\, f_j\\ &= f_i\, \bar{v}_j\, c_j - f_i\, v_i\, c_j\end{aligned} \tag{3.23}$$

where $\bar{v}$ is the average velocity defined by $c\bar{v} = \int dv\, vf$.

We now realize that the indices i and j both refer to the entire set of cars, so from Eq. (3.18) we have

$$\left(\frac{\partial f_i}{\partial t}\right)_{\text{int}} = \sum_j (\Gamma_{ij}^{(+)} - \Gamma_{ij}^{(-)})\,(1-P)$$
$$= f_i(x, v_i, t)\, c(x, t)\, [\bar{v}(x, t) - v_i]\,(1-P). \tag{3.24}$$

Using (3.17) and (3.24), our kinetic equation (3.16) for the evolution of the velocity distribution function can be written as

$$\frac{\partial f}{\partial t} + v\frac{\partial f}{\partial x} = -\frac{f-f_0}{T} + (1-P)\,c(\bar{v}-v)f. \tag{3.25}$$

This is a nonlinear integro-differential equation in the unknown distribution function in which $\bar{v}$ and c depend implicitly on f. Most of the work described in what follows is based on this equation.

It is of interest to check that Eq. (3.25) preserves the normalization of f in time and, therefore, does not conflict with the conservation equation (2.1). Indeed, if we integrate (3.25) over all velocities, we obtain

$$\frac{\partial c}{\partial t} + \frac{\partial(\bar{v}c)}{\partial x} = -\frac{1}{T}\int dv\,(f-f_0) + (1-P)c[\bar{v}\textstyle\int dv\, f - \int dv\, vf] = 0. \tag{3.26}$$

Each of the two terms on the right-hand side in the foregoing equation vanishes separately, and (2.1) is recovered.

Qualitatively, the interaction term in (3.25) has exactly the effect we would expect: for $v > \bar{v}$ it decreases the value of f, whereas for $v < \bar{v}$ it increases f. We may verify this conclusion by deriving from (3.25) the equation for the time evolution of the average velocity. We multiply (3.25) by v and integrate. For homogeneous conditions (i.e., c = constant) we find

$$\frac{d\bar{v}}{dt} = -\frac{\bar{v}-\bar{v}^0}{T} + (1-P)\,c[(\bar{v}^2) - \overline{v^2}] \tag{3.27}$$

where $\bar{v}^0$ is the mean desired speed.

This gives the time-independent steady-state solution

$$\bar{v}(\infty) = \bar{v}^0 + T(1-P)\,c[(\bar{v}(\infty))^2 - \overline{v^2(\infty)}] \tag{3.28}$$

and we obtain, remembering that

$$\overline{(v-\bar{v})^2} = \overline{v^2} - (\bar{v})^2 > 0,$$

the following:

$$\bar{v}(\infty) < \bar{v}_0 \,. \tag{3.29}$$

The average velocity is, therefore, lowered, which is a very reasonable conclusion. In the derivation of the kinetic equation, correlations between cars have been neglected. For example, in Eq. (3.19) we have the product of two one-car distribution functions. It would be more correct to introduce a two-car distribution function f_2 to describe such situations (Prigogine and Andrews, 1960). This is the standard procedure in statistical physics. The difference between f_2 and the product of this one-car distribution function measures the correlations between the two interacting cars.

It may be worthwhile to follow up this possibility in greater detail, especially for higher-density traffic. However, it should be emphasized again that many gross features of multilane traffic are still rather badly understood (i.e., the modification of the desired velocity distribution with concentration, Section 3.2). At present it seems uncertain that better agreement with data, when available, will be achieved in this way.

It should be pointed out that the probability of "non-passing", $1-P$, as it appears in the interaction term (3.18) refers to a description of the highway in which, as we have mentioned, correlations have been neglected. The probability of non-passing is therefore itself an average quantity taken over a sufficiently large sample. Here again it is possible to suggest improvements such as velocity dependent passing probabilities or correlations between the local state and the passing probability. In the latter case P would become a functional of f and not only of c. In the absence of experimental information such improvements seem to us to be premature. If necessary, they can be introduced at a later stage.

It has been stated (Munjal and Pahl, 1969) that the interaction term in (3.25) is limited to traffic situations where no vehicles are queueing. This statement is wrong. The distribution function f refers to a state of the road containing a large number of cars in all possible situations (including queueing). We have already emphasized this condition in Section 2.1. It is precisely the occurrence of hindrance mechanisms like queueing that is responsible for the alteration of the distribution function, that is, the decrease in the average velocity: see (3.29). The mistake made by Munjal and Pahl (1969, p. 157) is to assume that a queue of n vehicles with speed v and position x

would be represented in our theory by a value of $f(x, v, t)$ that would be n times as high as for the case of one vehicle with the same x and v. This assumption confuses the density of a gas with the density of a molecule! Unfortunately, such misconceptions may mislead readers without the necessary background in physics (Prigogine and Herman, 1970).

In conclusion, it seems reasonable to consider Eq. (3.25) as a model equation for highway traffic. This equation plays the same central role in our theory as the fundamental Boltzmann equation in the kinetic theory of gases. This is the reason we call our approach the Boltzmann-type approach to traffic theory. As will become clear in the following chapters, it is indeed possible to describe the essential broad features of traffic as changes in flow behavior, instabilities, and so forth; moreover, in spite of its simple appearance, the traffic kinetic equation leads to interesting and new problems.

3.5 Structure of the Kinetic Equation and Invariants

In the following chapters we will discuss the solution of the kinetic Eq. (3.25); here we first present a few general remarks (Prigogine *et al.*, 1967). We write Eq. (3.16) formally as

$$\frac{df}{dt} = O(f). \tag{3.30}$$

The functional $O(f)$ has to express the competition between the desires of the individual drivers and their mutual interactions.

It is, therefore, very natural to assume that $O(f)$ may be written as a sum of a linear function, say $\mathscr{D}(f)$, corresponding to an "independent" behavior of the drivers and a nonlinear function, say $\mathscr{I}(f)$ corresponding to various interaction processes. Thus,

$$O(f) = \mathscr{D}(f)+\mathscr{I}(f). \tag{3.31}$$

Equation (3.25) is precisely of this type. The only general restriction that has to be imposed on the functional $O(f)$ is the preservation of the total number of cars. Therefore, it has to satisfy the identity

$$\int_0^\infty dv\, O(f) = 0. \tag{3.32}$$

As we have seen, this condition is fulfilled by Eq. (3.25). As indicated in Chapter 2, however, there is no reason to expect that characteristics of the car population, other than their number, should be conserved. The conservation of energy and momentum can be ruled out in the general case because of the existence of the well-known situation in which a fast car closing in on a slow car is unable to pass and adopts the speed of the slow car; neither energy nor momentum is conserved in such a process.

We therefore assume that (3.32) is the *only integral identity* imposed on $O(f)$. This is in sharp contrast with the kinetic gas equation (i.e., the Boltzmann equation)

$$\frac{df}{dt} = \frac{\partial f}{\partial t} + v\frac{\partial f}{\partial x} + \mathscr{F}\frac{\partial f}{\partial v} = \mathscr{C}(f). \tag{3.33}$$

Here $f(x, v, t)$ is the velocity distribution of the molecules, $\mathscr{F}$ the external force, and $\mathscr{C}(f)$ the usual collision operator. The main difference between the usual Boltzmann equation and the traffic Boltzmann equation (3.25) is that we have in the former case, of course, no desired distribution function, and therefore no relaxation term.

In the gas kinetic case, the collision operator $\mathscr{C}(f)$ now satisfies *five* conditions (conservation of number, momentum, and energy through collisions):

$$\int d\mathbf{v}\,\mathscr{C}(f) = 0, \tag{3.34a}$$

$$\int d\mathbf{v}\,\mathbf{v}\mathscr{C}(f) = 0, \tag{3.34b}$$

and

$$\int d\mathbf{v}\,v^2\mathscr{C}(f) = 0. \tag{3.34c}$$

As a consequence, the most general equilibrium solution of (3.33) describing a homogeneous time-independent system contains the five conserved quantities

$$c, \qquad \bar{\mathbf{v}}, \qquad \mathrm{T} \tag{3.35}$$

corresponding respectively to the concentration

$$\int d\mathbf{v}\,f = c, \tag{3.36}$$

the average velocity

$$\frac{1}{c}\int d\mathbf{v}\,\mathbf{v}f = \bar{\mathbf{v}}, \tag{3.37}$$

and the temperature

$$\tfrac{1}{2}\int d\mathbf{v}\,m(\mathbf{v}-\bar{\mathbf{v}})^2 f = \tfrac{3}{2}k\mathrm{T}. \tag{3.38}$$

In order to derive the hydrodynamic description (Euler or Stokes–Navier equation) from the kinetic equation, we require the assumption that the system is in "local equilibrium." By local equilibrium is meant that the quantities that characterize the system are varying sufficiently slowly with respect to space and time so that the same description holds as in the homogeneous time-independent case but with the local values of the quantities. More precisely the most general local equilibrium solution of (3.33) now contains the five functions

$$c(\mathbf{x}, t), \qquad \bar{\mathbf{v}}(\mathbf{x}, t), \qquad \mathrm{T}(\mathbf{x}, t). \tag{3.39}$$

In other words, we may in the gas kinetic case prescribe independently five moments of the distribution function, whereas in the traffic flow case we may only prescribe the single moment (1.2).

The local equilibrium solutions of kinetic theory are of the form

$$f = f[c(\mathbf{x}, t), \bar{\mathbf{v}}(\mathbf{x}, t), \mathrm{T}(\mathbf{x}, t), \mathbf{v}]; \tag{3.40}$$

the corresponding solutions in the traffic case are of the form

$$f = f[c(x, t), v] \tag{3.41}$$

in which case the space-time behavior of f appears only through the local concentration. As a consequence of (3.41), the local average velocity becomes a functional of $c(x, t)$, namely,

$$\bar{v}(x, t) = \frac{1}{c(x, t)} \int_0^\infty dv\, v f[c(x, t), v] = \bar{v}[c(x, t)]. \tag{3.42}$$

This is precisely the basic assumption of the hydrodynamical theory we summarized in Chapter 2 (see (2.7)). We will discuss its validity in detail in Chapter 5. In the present section we want only to stress the close relationship between the form of traffic hydrodynamics and the structure of the traffic kinetic equation, as given by (3.25).

3.6 Adjustments

More general expressions for the interaction term in (3.17) or (3.31) may be considered. For example, effects of the type introduced by the follow-the-leader approach may also be incorporated (for more details see Prigogine *et al.*, 1962b). As we have seen, typical laws expressing the follow-the-leader theory are of the form (see Section 1.2)

$$\frac{du_n}{dt} = \lambda(u_{n-1} - u_n) \tag{3.43}$$

where u_n is the speed of the nth car. In (3.43) λ may be a function of the relative distance between the $(n-1)$st and nth cars. In this equation we have neglected the effect of time lags. An important factor is that every expression of the form

$$u_n = \phi, \tag{3.44}$$

where ϕ is an arbitrary constant independent of n, is a solution of Eq. (3.43) in the time independent case. The main physical consequence of the adjustment expressed by Eq. (3.43) is therefore a trend towards a smaller dispersion in the speed distribution.

A qualitatively similar effect can be achieved by adding to the kinetic equation a new term of the general form

$$\left(\frac{\partial f}{\partial t}\right)_{\text{adj}} = \lambda(c)\,(1-P)\,c[c\delta(v-\bar{v})-f] \tag{3.45}$$

where as usual $\delta(\xi)$ signifies the Dirac delta-function. (Because of the similar qualitative meaning we have used the symbol λ in both Eq. (3.43) and Eq. (3.45). In (3.45) λ may be an arbitrary function of concentration.) Equation (3.45) satisfies the following conditions:

$$\int_0^\infty dv \left(\frac{\partial f}{\partial t}\right)_{\text{adj}} = \lambda(1-P)\,c\left[c\int_0^\infty dv\ \delta(v-\bar{v}) - c\right] = 0. \tag{3.46}$$

Therefore, using (3.17) we have

$$\frac{dc}{dt} = \int_0^\infty dv \left(\frac{\partial f}{\partial t}\right)_{\text{rel}} + \int_0^\infty dv \left(\frac{\partial f}{\partial t}\right)_{\text{int}} + 0. \tag{3.47}$$

The time evolution of the concentration is not altered by the adjustment term. Similarly

$$\int_0^\infty dv \left(\frac{\partial f}{\partial t}\right)_{\text{adj}} v = \lambda(1-P)\,c\left[c\int_0^\infty dv\ v\delta(v-\bar{v}) - \bar{v}c\right] = 0 \tag{3.48}$$

and

$$\frac{d(c\bar{v})}{dt} = \int_0^\infty dv \left(\frac{\partial f}{\partial t}\right)_{\text{rel}} v + \int_0^\infty dv \left(\frac{\partial f}{\partial t}\right)_{\text{int}} v + 0. \tag{3.49}$$

For homogeneous situations (c = constant) Eqs. (3.22) and (3.28) still remain valid, and the average velocity still satisfies Eq. (3.28) for long times. (In the usual terminology of the kinetic theory of gases, (3.48) expresses the "conservation of velocity" in the adjustment process. This does *not* mean that the average velocity will be unaffected. According to (3.28), the average velocity depends on the fluctuations, which themselves are decreased by the adjustment process.) Finally we have

$$\begin{aligned}\int_0^\infty dv \left(\frac{\partial f}{\partial t}\right)_{\text{adj}} (v-\bar{v})^2 &= \lambda(1-P)\, c\left[c\int_0^\infty dv\, (v-\bar{v})^2\, \delta(v-\bar{v}) - c\overline{(v-\bar{v})^2}\right] \\ &= \lambda(1-P)\, c^2\overline{(v-\bar{v})^2} \le 0. \end{aligned} \tag{3.50}$$

Therefore the adjustment term will tend to decrease the dispersion. Needless to say, this will not be the "only" effect. Every modification of the kinetic equation changes the evolution and the steady state of the distribution function and therefore of all moments (see Munjal and Pahl, 1969, p. 160).

Other more complicated choices could be considered but are not very likely to change anything in our qualitative conclusions. We can therefore express the follow-the-leader effect by adding to the Boltzmann-like equation the term given in (3.45). In this equation $\lambda(c)$ is a still-unspecified function of the concentration and therefore of the average distance between the drivers; we also multiply $\lambda(c)$ by $(1-P)$ because the follow-the-leader effects must disappear in the limit of free passing or $P = 1$.

The term in Eq. (3.45) has a very simple intuitive meaning. It expresses that each driver is subjected to a kind of collective effect, exerted by the total ensemble on each driver, which tends to increase the speed of the slow cars and to decrease the speed of the rapid cars. The effect on the rapid cars is similar to that of the interaction term, while the effect on the slow cars is opposite to the interaction term.

The relaxation term is proportional to f_0-f; the adjustment term has a similar form, as it is proportional to $c\delta(v-v)-f$. It is therefore obvious that we may combine these two mechanisms and write (Munjal and Pahl, 1969)

$$\begin{aligned}\left(\frac{\partial f}{\partial t}\right)_{\text{rel}} + \left(\frac{\partial f}{\partial t}\right)_{\text{adj}} &= \frac{f_0-f}{T} + \beta[c\delta(v-\bar{v})-f] \\ &= \frac{f^*-f}{T'} \end{aligned} \tag{3.51}$$

where

$$f^* = \xi f_0^* + (1-\xi)c\delta(v-\bar{v}), \qquad \xi = \frac{T'}{T},$$

$$\beta = \lambda(1-P)c, \qquad T' = \frac{T}{(1+\beta T)}.$$

The change from f_0 to f^* may be considered as a special case of adapting a new "strategy" depending on the conditions that prevail. This will bc studied in more detail in Chapter 6. When such adjustments are taken into account, Eq. (3.12) is no longer valid, since the effective desired distribution function then depends on the concentration in a more complicated manner.

The individual parameters in Eq. (3.51), such as ξ, β, and T', are very difficult to determine from experimental or observational data. However, these parameters can be determined implicitly from a pair of observed speed distributions, such as f_0 and f or, in principle, any pair. Recently, such an approach has been used by Gafarian *et al.* (1970) in an attempt to verify Eq. (3.51) with respect to observational data obtained on a multilane highway. It is regrettable that their verification techniques contain undesirable features that render their results inconclusive with respect to the assumptions and properties of the model. It should also be mentioned that Kraus (1966) attempted to compare our kinetic theory model with observational data. He stated that the tests were too limited to be conclusive.

3.7 Appendix

Going back to Eq. (3.15)

$$\frac{\partial f}{\partial t} + v\frac{\partial f}{\partial x} = -\frac{f-f_0}{T}, \tag{3A.1}$$

where T is regarded as a constant relaxation time, we are interested in the time evolution of the initial distribution $f(x, v; t = 0)$ to some steady-state solution $f(x, v, t)$. This behavior can be examined by using a Fourier–Laplace transform technique for which the poles yield the required time behavior. The Fourier–Laplace transform of (3A.1) is given by

$$-f_k(v; 0) - iz\varrho_k(v; z) + vik\varrho_k(v; z) = \frac{-\varrho_k(v; z) + f_{0,k}(v; z)}{T}, \tag{3A.2}$$

or, after rearranging terms, by

$$\varrho_k(v;z) = \frac{f_k(v;0)+f_{0,k}(v;z)/T}{ikv-iz+T^{-1}}, \tag{3A.3}$$

where

$$f(x,v;t) = \frac{1}{2\pi}\int dz\, e^{-izt}\int_{-\infty}^{+\infty} dk\, e^{ikx}\varrho_k(v;z),$$

$$f(x,v;0) = \int dk\, e^{ikx} f_k(v;0), \tag{3A.4}$$

$$f_0(x,v,t) = \frac{1}{2\pi}\int dz\, e^{-izt}\int dk\, e^{ikx} f_{0,k}(v;z).$$

Note that transformed variable iz used here is usually denoted by $-s$ in the literature on Laplace transforms.

We introduce the probability distribution $\tilde{f}_0(v)$ related to f_0 through (3.12):

$$f_0(x,v;t) = c(x,t)\tilde{f}_0(v) \tag{3A.5}$$

with

$$\int_0^\infty dv\, \tilde{f}_0(v) = 1. \tag{3A.6}$$

The Fourier–Laplace Transform of $c(x, t)$ is given by

$$c(x,t) = \frac{1}{2\pi}\int dz\, e^{-izt}\int dk\, e^{ikx}\nu_k(z). \tag{3A.7}$$

We now note that because of (3A.4) and (3A.7)

$$\int_0^\infty dv\, \varrho_k(v;z) = \nu_k(z). \tag{3A.8}$$

Hence, integration of (3A.3) yields

$$\nu_k(z) = \frac{\displaystyle\int_0^\infty dv\,\frac{f_k(v;0)}{ikv-iz+T^{-1}}}{\displaystyle 1-\frac{1}{T}\int_0^\infty dv\,\frac{f_0(v)}{ikv-iz+T^{-1}}}. \tag{3A.9}$$

Let us examine the behavior of the denominator for the case in which

$$f_0(v) = \frac{1}{v_{\max}}, \qquad 0 \le v \le v_{\max}, \tag{3A.10}$$

$$= 0, \qquad v_{\max} < v.$$

Then

$$1-\frac{1}{T}\int_0^\infty dv\,\frac{f_0(v)}{ikv-iz+T^{-1}} = 1-\frac{1}{ikv_{\max}T}\ln\left(\frac{ikv_{\max}-iz+T^{-1}}{-iz+T^{-1}}\right) \tag{3A.11}$$

and the zeros of (3A.11) all correspond to

$$\frac{kv_{\max}-z-iT^{-1}}{-z-iT^{-1}} = \exp\,(ikv_{\max}T). \tag{3A.12}$$

Setting the real and imaginary parts of (3A.12) separately equal to zero yields

$$\omega(1-\cos\chi)+\gamma\sin\chi = kv_{\max}-T^{-1}\sin\chi \tag{3A.13}$$

$$-\omega\sin\chi+\gamma(1-\cos\chi) = -T^{-1}(1-\cos\chi) \tag{3A.14}$$

where $\chi = kv_{\max}T$. We can then obtain the following expression for ω:

$$\omega = \frac{\begin{vmatrix} (kv_{\max}-T^{-1}\sin\chi) & \sin\chi \\ -T^{-1}(1-\cos\chi) & (1-\cos\chi) \end{vmatrix}}{\begin{vmatrix} (1-\cos\chi) & \sin\chi \\ -\sin\chi & (1-\cos\chi) \end{vmatrix}}. \tag{3A.15}$$

We also obtain

$$\begin{aligned}\gamma &= \frac{\begin{vmatrix} (1-\cos\chi) & (kv_{\max}-T^{-1}\sin\chi) \\ -\sin\chi & -T^{-1}(1-\cos\chi) \end{vmatrix}}{2(1-\cos\chi)} \\ &= -\frac{1}{T}+\tfrac{1}{2}kv_{\max}\cot\frac{\chi}{2}.\end{aligned} \tag{3A.16}$$

An important observation is the behavior of γ for small relaxation times (i.e., $T\sim 0$). From (3A.16) we find that

$$\begin{aligned}\gamma &= -\frac{1}{T}+\frac{kv_{\max}}{2}\left[\frac{\chi}{2}+\frac{1}{3}\left(\frac{\chi}{2}\right)^3+\ldots\right]^{-1} \\ &\cong -\frac{1}{T}+\frac{kv_{\max}}{\chi}\left[1+\frac{1}{3}\left(\frac{\chi}{2}\right)^2\right]^{-1} \\ &= \frac{1}{T}\left[-1+1-\frac{1}{3}\left(\frac{\chi}{2}\right)^2\right] \\ &= -\frac{1}{3}\left(\frac{kv_{\max}}{2}\right)^2 T.\end{aligned} \tag{3A.17}$$

Hence we have from (3A.17) that

$$\lim_{T\to 0} \gamma = 0. \tag{3A.18}$$

The important conclusion to be drawn from this calculation is contained in Eqs. (3A.17) and (3A.18) and involves a comparison of the time behavior of the contributions from the numerator of (3A.9) with those of the denominator of (3A.9). This can simply be illustrated with the example of an initial distribution $f(x, v, t = 0)$ that is a velocity spike at some speed u. Now the contributions from the numerator of (3A.9) for such a case would be damped with an exponential exp $(-t/T)$, and as $T\to 0$ this damping becomes very strong; whereas the contributions from the denominator are not damped at all in this limit, as indicated by (3A.18). Hence, we have two time scales appearing, one associated with the initial distribution and the other associated with the properties of the desired speed distribution.

The disappearance of the poles associated with the numerator of (3A.9) and the vanishing of the damping associated with the zeros of the numerator of (3A.9) for small T is a general result. It can be derived as follows:

$$\nu_k(z) = \frac{\displaystyle\int_0^\infty dv \, \frac{f_k(v;0)}{ikv - iz + T^{-1}}}{1 - \displaystyle\frac{1}{T}\int_0^\infty dv \, \frac{f_0(v)}{ikv - iz + T^{-1}}} \tag{3A.19}$$

$$= \frac{T\int_0^\infty dv\, f_k(v;0)\,\{1 - [(ikv - iz)T] + \frac{1}{2}[(ikv - iz)T]^2 + \cdots\}}{1 - \int_0^\infty dv\, f_0(v)\,\{1 - [(ikv - iz)T] + \frac{1}{2}[(ikv - iz)T]^2 + \cdots\}}.$$

Now introducing the notation

$$c_k(0) = \int_0^\infty dv\, f_k(v;0), \tag{3A.20}$$

we find that in the limit as $T\to 0$, Eq. (3A.19) becomes

$$\lim_{T\to 0} \nu_k(z) = \frac{c_k(0)}{ik\bar{v}^0 - iz}. \tag{3A.21}$$

Hence, the only singularities present are zeros of the denominator, as we asserted above. The short time scale corresponds to the "forgetting" of the initial velocity perturbation, while the other time scale refers to the slow evolution of the local concentration.

The Fourier–Laplace techniques we have employed here to discuss Eq. (3A.1) will be used in Chapter 5 to discuss the full kinetic equation. (3.25).

CHAPTER 4

Solution of the Kinetic Equation: Individual and Collective Flow

4.1 Kinetic Equation for Homogeneous Time-Independent Situations

We now turn our attention again to the kinetic equation (3.25)

$$\frac{\partial f}{\partial t} + v\frac{\partial f}{\partial x} = -\frac{(f - f_0)}{T} + c(\bar{v} - v)(1 - P)f \tag{4.1}$$

where f_0 = one-car desired speed distribution function, T = relaxation time, $c = \int_0^\infty dv\, f = \int_0^\infty dv\, f_0$ = concentration, $\bar{v} = c^{-1} \int_0^\infty dv\, vf$ = average speed, and P = probability of passing.

As we have seen, the quantity f_0 is a distribution function that characterizes the driver–car–road complex in the absence of interaction between the various cars. Therefore, f_0 is considered to be that distribution function which would be achieved for infinite dilution. The function f_0 can incorporate into the theory such information as the wishes of the drivers, response of the driver–car system, speed limits, and the characteristics of the road.

The term $c(\bar{v} - v)(1 - P)f$ represents the net change in f caused by interaction processes in which cars, when interacting with cars ahead moving with slower speeds, either pass these cars or assume the slower speeds.

As can be seen from (4.1), the homogeneous time-independent (i.e., stationary) solution f satisfies the nonlinear equation

$$f = \frac{f_0}{1 - cT(1 - P)(\bar{v} - v)} \tag{4.2}$$

where by homogeneous we mean that f is not spatially dependent. The quantity f describes the situation in which there is a steady state between the slowing down of cars caused by interaction processes and the speeding up of cars caused by passing. To perform actual calculations starting with (4.1) we need to know how the relaxation time T and the probability of passing P vary with the concentration. Many different mechanisms may contribute to T. For example, by the definition of f_0, a car moving from a position where f_0 has one value to another where the value of f_0 is different will achieve the new desired distribution in some time T_1. Another relaxation mechanism will be involved in passing, and this will correspond to some other relaxation time T_2. We can expect the relaxation time T_2 to be of the form

$$T_2 = \frac{\tau(1-P)}{P} \,. \tag{4.3}$$

The total relaxation time T will be dominated by the larger of the two relaxation times involved. We consider here a situation in which

$$T_2 \gg T_1 \tag{4.4}$$

so that

$$T \simeq T_2 = \frac{\tau(1-P)}{P} \,. \tag{4.5}$$

We have next to consider the probability of passing. It is assumed that the probability of passing takes the explicit form

$$P = 1-c/c_P = (1-\eta), \qquad c < c_P \tag{4.6}$$

and

$$P = 0, \qquad c > c_P,$$

where

$$\eta = \frac{c}{c_P}$$

and c_P is the limiting or jam concentration on an n-lane highway when the cars can no longer pass. It is of interest to note that the subsequent derivation of f depends, apart from the usual assumption that f is a well-behaved function, only on the assumption that T is independent of the speed v and not on the particular forms given in Eqs. (4.5) and (4.6). However, since all the numerical work presented in this chapter is based on these forms, they are introduced throughout the following treatment. Note that for small concentrations T is quite small, whereas for large concentrations it is very large, which qualitatively must be the case whatever the explicit form of T. Hence, these particular forms do not appear unduly restrictive. Let us now discuss in detail the solution of Eq. (4.2).

4.2 Individual and Collective Flow

As indicated by Eq. (4.2), we have to distinguish between two cases (Prigogine *et al.*, 1962a).

Case A. If we consider the case in which

$$1 - Tc(1-P)\bar{v} > 0, \tag{4.7}$$

then f is given by Eq. (4.2) as follows:

$$f = \frac{f_0}{1+Tc(1-P)(v-\bar{v})}. \tag{4.8}$$

This solution reduces to the ideal or desired speed distribution function in the limit of vanishing concentration. However, it is clear that this solution cannot, in general, be valid for arbitrarily high concentrations because whenever

$$1 - Tc(1-P)\bar{v} < 0, \tag{4.9}$$

the distribution function may become negative, which is clearly absurd.

Case B. In this case we consider

$$1 - Tc(1-P)\bar{v} = 0. \tag{4.10}$$

Then Eq. (4.2) reduces to

$$f = \frac{f_0}{Tc(1-P)v}. \tag{4.11}$$

The important feature to be noted here is that the homogeneous equation corresponding to Eq. (4.11), namely,

$$Tc(1-P)vf = 0, \tag{4.12}$$

admits the singular solution

$$f = \alpha c \delta(v) \tag{4.13}$$

where α is an arbitrary constant and $\delta(v)$ is the Dirac delta function. Therefore, the general solution of Eq. (4.11) is of the form

$$f = \frac{f_0}{Tc(1-P)v} + \alpha c \delta(v). \tag{4.14}$$

The solution given in Eq. (4.8) corresponds to what may be called the *individual flow pattern* and is related in a simple way to the ideal or desired speed distribution function.

The second solution corresponds, on the other hand, to what may be called the *collective flow pattern.* Indeed, as shown by Eq. (4.10), the average speed then depends only on the concentration, the probability of passing P, and the relaxation time T (both P and T are themselves functions of the concentration), and is independent of the desired speed distribution function and, therefore, of the wishes of the drivers. This solution is characterized by the occurrence of a singularity at the origin. However, the critical concentration at which the individual flow becomes collective does depend on the desired speed distribution.

In both cases the time-dependent solution may be written in the following form:

$$f = \frac{f_0}{1+Tc(1-P)(v-\bar{v})} + \alpha c\delta(v) \tag{4.15}$$

where α is an undetermined constant that may be identically zero. This solution has to satisfy the following two requirements:

A. *Normalization.* This gives us the relation

$$c = \int_0^\infty \frac{dv\, f_0}{1+Tc(1-P)(v-\bar{v})} + \alpha c. \tag{4.16}$$

B. Average Speed. Multiplying Eq. (4.15) by v and integrating, we obtain the condition

$$c\bar{v} = \int_0^\infty \frac{dv\, vf_0}{1+Tc(1-P)(v-\bar{v})}. \tag{4.17}$$

Equation (4.17) with the aid of the normalization condition Eq. (4.16) may be transformed into

$$\alpha[1-Tc(1-P)\bar{v}] = 0. \tag{4.18}$$

We see, therefore, that we have two solutions. The first solution corresponds to $\alpha = 0$ (i.e., individual flow), whereas the second corresponds to the case

$$Tc(1-P)\bar{v} = 1 \tag{4.19}$$

with α different from zero (i.e., collective flow).

There exists a striking analogy in the mathematical situation described by our equations with the so-called Bose–Einstein condensation (see, e.g., Tolman, 1938). In that case, above a certain concentration (for a given temperature) the population distribution splits into two parts, a regular part and a singular one. If the concentration is further increased beyond its critical value, the singular part increases relative to the regular part. However, the Bose–Einstein condensation occurs for a system in thermodynamic equilibrium, whereas the situation described here is closer to a nonequilibrium stationary state. This situation is similar to a time-independent heat conduction problem.

Returning to the traffic situation, the speed distribution above the critical concentration is characterized by large fluctuations indicating, broadly speaking, local traffic jams. It should be observed that the general solution of Eq. (4.2) is of the form

$$f = \frac{f_0}{1+Tc(1-P)(v-\bar{v})} + \alpha c\delta[1+Tc(1-P)(v-\bar{v})]. \qquad (4.20)$$

However, for Case A above, that is,

$$1-Tc(1-P)\bar{v} > 0, \qquad (4.21)$$

the corresponding singularity occurs for a negative value of v and we have therefore to take $\alpha = 0$. On the contrary, if we could have

$$1-Tc(1-P)\bar{v} < 0, \qquad (4.22)$$

then the singularity would occur for positive values of v. Then, however, the regular part of Eq. (4.20), and therefore also f itself, would become negative for some small values of v, which is impossible if we assume that f_0 does not vanish for these values of v. Therefore, we have only the two possibilities given by Eqs. (4.7) and (4.10) corresponding to individual or collective flow, respectively.

4.3 Low-Density Traffic

Let us first consider "dilute" systems, that is, concentrations much smaller than the critical concentration c_{crit} (see 4.10) given by

$$c_{\text{crit}} = [\bar{v}T(1-P)]^{-1} \qquad (4.23)$$

at which the transition to collective flow occurs. We then obtain from (4.8) as the first correction to the desired speed distribution function

$$f = f_0[1 - Tc(1-P)(v - \bar{v}^0)] \tag{4.24}$$

or

$$f = f_0[1 - \gamma(v - \bar{v}^0)] \tag{4.25}$$

with

$$\gamma = Tc(1-P). \tag{4.26}$$

Note that $\bar{v}^0$ denotes that the average is taken using the desired speed distribution f_0.

If the explicit expressions (4.5) and (4.6) are employed, we may also write (4.26) in the form

$$\gamma = c_P T \frac{\eta^3}{1-\eta}. \tag{4.27}$$

For dilute systems γ varies approximately as the third power of the concentration. In terms of a parameter λ (not to be confused with the adjustment parameter), defined as

$$\lambda = 1 - \gamma\bar{v}, \tag{4.28}$$

the condition (4.10) for collective flow becomes

$$\lambda = 0. \tag{4.29}$$

Using (4.25) we may calculate the form of the flow at low concentration. We obtain

$$q = c[\bar{v}^0 - \gamma(\overline{v^2}^0 - (\bar{v}^0)^2)]. \tag{4.30}$$

The result is interesting as it shows that the first order deviations from the linear portion of the flow curve are determined by the dispersion of the desired speed distribution function. For higher concentrations other moments such as the skewness contribute to this effect.

Also for higher concentrations it appears impossible to express the distribution function f in closed form. Therefore, numerical computations become necessary. The results of some of these calculations are presented in the next section.

4.4 Numerical Solution of the Kinetic Equation

Numerical results have been obtained for six different desired speed distribution functions (Herman *et al.*, 1962). Some of the properties of these models are given in Table 4.1. Note that the

Table 4.1. Properties of Various Models[a]

Free-speed distribution function models	f_0	$\bar{v}^0$	f	λ
1. Exponential	$\left(\frac{c_P\eta}{v_0}\right)\exp\left(\frac{-v}{v_0}\right)$	v_0	$\frac{(c_P\eta/v_0)[\exp(-v/v_0)]}{\lambda+\gamma v}$	$(\Gamma v_0)^{-1}\exp(\Gamma v_0)^{-1} \times \int_{(\Gamma v_0)^{-1}}^{\infty} dx\, x^{-1}\exp(-x)$
2. Modified exponential	$\left(\frac{4c_P\eta v}{v_0^2}\right)\exp\left(\frac{-2v}{v_0}\right)$	v_0	$\frac{(4c_P\eta v/v_0^2)[\exp(-2v/v_0)]}{\lambda+\gamma v}$	$2(\Gamma v_0)^{-1}\left\{1-2(\Gamma v_0)^{-1}\left[\exp\left(\frac{2}{\Gamma v_0}\right)\right] \int_{2(\Gamma v_0)^{-1}}^{\infty} dx\, x^{-1}\exp(-x)\right\}$
3. Straight line with a negative slope	$(2c_P\eta/v_L)(1-v/v_L),\ 0 \le v \le v_L$; $0,\ v_L \le v$	$\frac{2}{3}(1-v_L)$	$\frac{(2c_P\eta/v_L)(1-v/v_L)}{\lambda+\gamma v}$	$2(\Gamma v_L)^{-1}[\ln(1+\Gamma v_L)-1 + (\Gamma v_L)^{-1}\ln(1+\Gamma v_L)]$
4. Constant	$c_P\eta/v_L,\ 0 \le v \le v_L$; $0,\ v_L \le v$	$\frac{v_L}{2}$	$\frac{(c_P\eta/v_L)}{\lambda+\gamma v}$	$(\Gamma v_L)^{-1}\ln(1+\Gamma v_L)$
5. Straight line with a positive slope	$2c_P\eta v/v_L^2,\ 0 \le v \le v_L$; $0,\ v_L \le v$	$\frac{2}{3}v_L$	$\frac{(2c_P\eta v/v_L^2)}{\lambda+\gamma v}$	$2(\Gamma v_L)^{-1}[1-(\Gamma v_L)^{-1} \ln(1+\Gamma v_L)]$
6. Parabolic	$3c_P\eta v^2/v_L^3,\ 0 \le v \le v_L$; $0,\ v_L \le v$	$\frac{3}{4}v_L$	$\frac{(3c_P\eta v^2/v_L^3)}{\lambda+\gamma v}$	$3(\Gamma v_L)^{-1}[\frac{1}{2}-(\Gamma v_L)^{-1} + (\Gamma v_L)^{-2}\ln(1+\Gamma v_L)]$

[a] Each entry for λ is a transcendental equation for λ since Γ itself is a function of λ, namely, $\Gamma = \gamma/\lambda$, $\gamma = c_P\tau\eta^3/(1-\eta)$. Once f_0 is known, the stationary solution f is given by $f = f_0/(\lambda+\gamma v)$ where $\lambda = 1-\gamma\bar{v}$. The quantity λ is a constant for a given concentration and is determined by normalizing f to the concentration c, i.e., $c = \int_0^\infty dv\, f/(\lambda+\gamma v)$. In principle all the important quantities may be computed, e.g., the flow q, which is given by $q = c\bar{v} = \int_0^\infty dv\, vf = c_P\eta(1-\lambda)/\gamma$.

definitions of the symbols are given in the table footnote, as well as in the preceding section. Various results concerning four of the models are presented in Figs. 4.1 through 4.8, and it is felt that these are sufficient to bring out the salient points investigated to date.

Figure 4.1 shows the normalized flow q/c_P as a function of the normalized concentration η for Model 1 (exponential: see Table 4.1). The family of curves displayed are for the case $c_P = 0.1$ sec/ft and an average desired speed $\bar{v}^0$ ranging from 5 ft/sec to 100 ft/sec. The linear behavior of the curves for low concentrations, which is common to all models, is evident (see Eq. (4.30)).

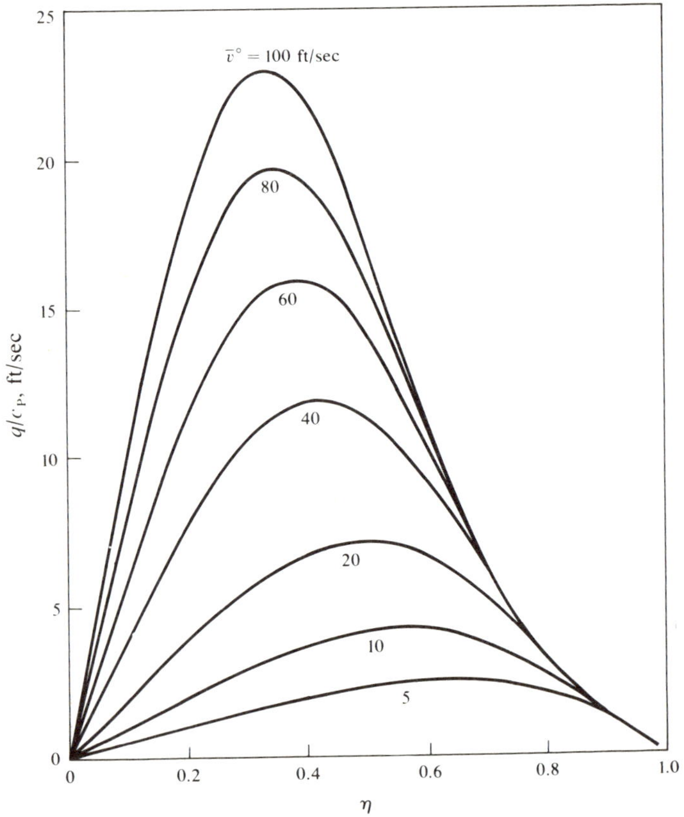

FIG. 4.1. Normalized flow q/c_P versus normalized concentration, $\eta = c/c_P$ for Model 1 (exponential: see Table 4.1) for different average desired speeds and $c_{PT} = 0.1$ sec/ft.

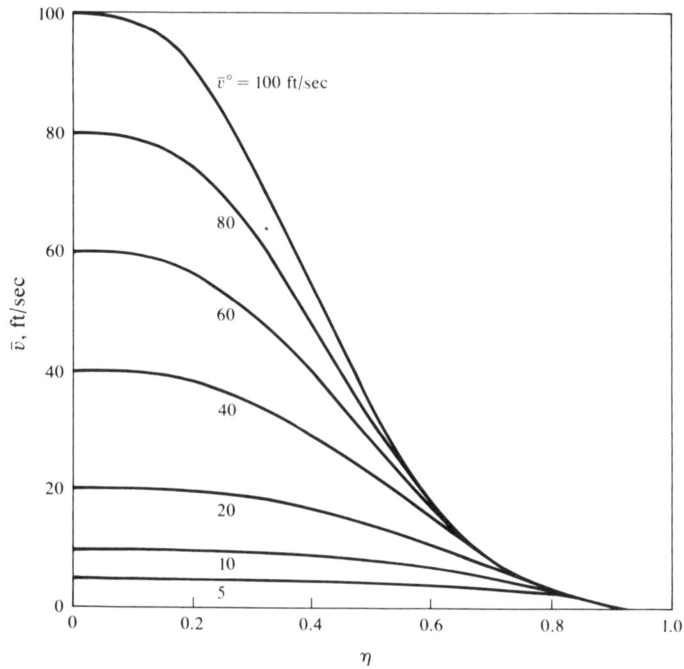

FIG. 4.2. Concentration dependence of the average speed for Model 1 (exponential: see Table 4.1) for different average desired speeds, $c_P =$ 0.05 car/ft, and $c_{PT} = 0.1$ sec/ft.

For the present model, as for all others investigated, the higher the average speed, the lower the concentration at which the optimum flow takes place. Note also that for $\eta \sim 1$ there are no essential differences in the flow for different average desired speeds, whereas for lower η there are dramatic differences. In Fig. 4.2 we have given the concentration dependence on the average speed for the exponential model. Figure 4.3 shows the normalized distribution function plotted versus speed. It illustrates how rapidly the distribution function changes with concentration.

Figure 4.4 shows the normalized flow as a function of the normalized concentration for Model 2. The family of curves displayed are for the same values of the parameters as employed for Model 1 in Fig. 4.1.

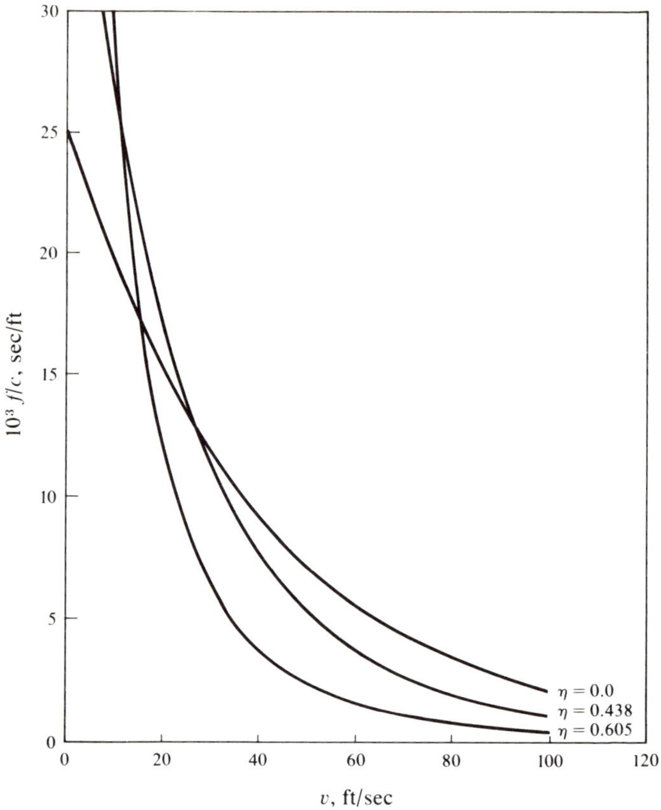

FIG. 4.3. Speed dependence of the normalized distribution function for Model 1 (exponential: see Table 4.1) for various concentrations, an average desired speed of 40 ft/sec, $c_P = 0.05$ car/ft, and $c_P\tau = 0.1$ sec/ft.

Figures 4.5 and 4.6 illustrate the individual and collective flow-concentration relationships for the modified exponential (infinite speed range) and the parabolic desired speed distribution functions, respectively. For a given form of the probability of passing P and given c_P and τ, the normalized concentration at which the individual flow curve intersects the collective flow curve η_{int} is dependent on the particular form taken for the desired speed distribution function f_0 and the average desired speed $\bar{v}^0$.

The intersection between the two curves can occur before the maximum of an individual flow-concentration curve is reached

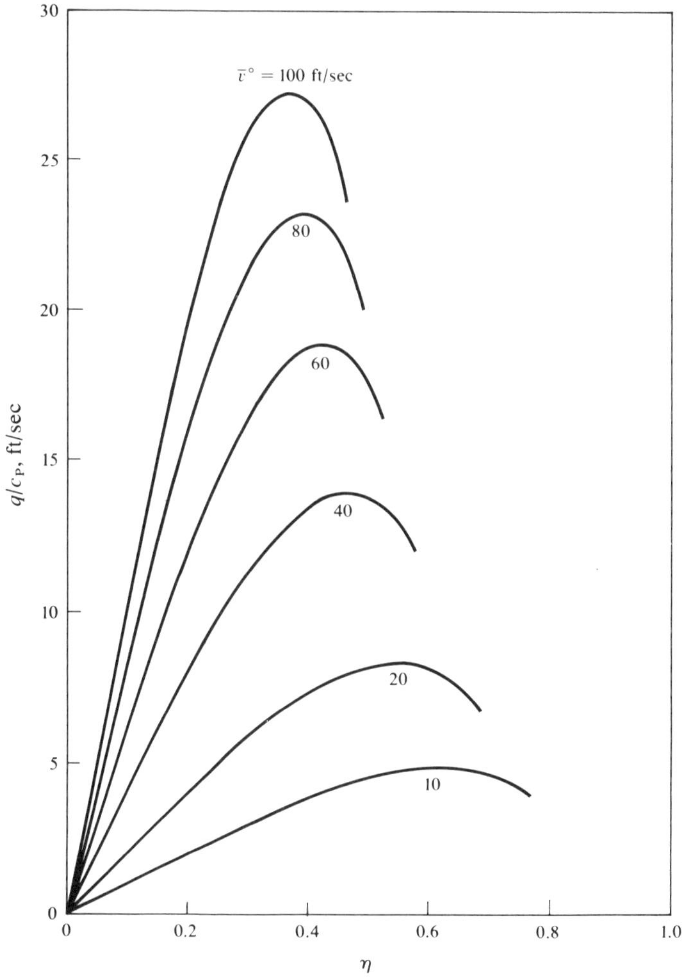

FIG. 4.4. Normalized flow q/c_P versus normalized concentration for the modified exponential desired speed distribution function, Model 2 in Table 4.1. Curves are given for different average desired speeds and $c_{PT} = 0.1$ sec/ft.

(see Fig. 4.6). This is so in the case of models that yield a small dispersion of the desired distribution function, and for high average desired speeds $\bar{v}^0$.

The amount of the deviation from the linear flow-concentration

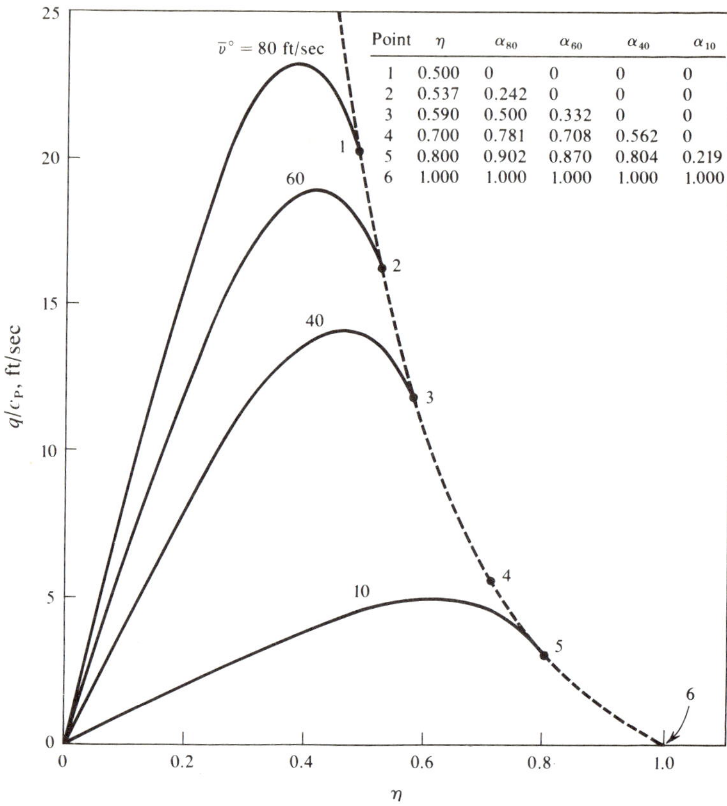

FIG. 4.5. Normalized flow q/c_P versus normalized concentration, $n = c/c_P$ for the modified exponential desired speed distribution function, Model 2 in Table 4.1. The solid curves represent the individual flow curves for various values of $\bar{v}^0$. The dashed curve represents the collective flow curve. The values of α associated with the irregular part of the achieved distribution function are given at several points of the collective flow curve for various values of $\bar{v}^0$. The probability of passing is taken as $P = 1 - \eta$ and $c_{PT} = 0.1$ sec/ft.

curve for a given $\bar{v}^0$ is determined by the dispersion associated with the desired speed distribution function f_0. A measure of this dispersion is the fluctuation $\overline{(v-\bar{v}^0)^2}^0$. Hence, we expect and find that there is a greater deviation for a given $\bar{v}^0$ for the modified exponential (infinite speed range) than for the parabolic desired speed distribution function. Although the smaller dispersion results in a smaller deviation, it also results in the individual flow characteristics terminating for relatively lower concentrations, as evidenced by noting that η_{int} (parabolic) is less than η_{int} (modified exponential) for the same $\bar{v}^0$. Beyond η_{int} the normalized flow is given by (4.10). This may be written

$$\frac{q}{c_P} = \frac{1}{c_P T(1-P)} \tag{4.31}$$

and is independent of f_0 or the desires of the drivers. For the collective flow state, however, the speed distribution function remains dependent on f_0. Equation (4.31) shows that the collective flow is dependent on the relaxation time T and the form chosen for the probability of passing P.

The speed dependence of the regular part of the achieved speed distribution function for the modified exponential (infinite speed

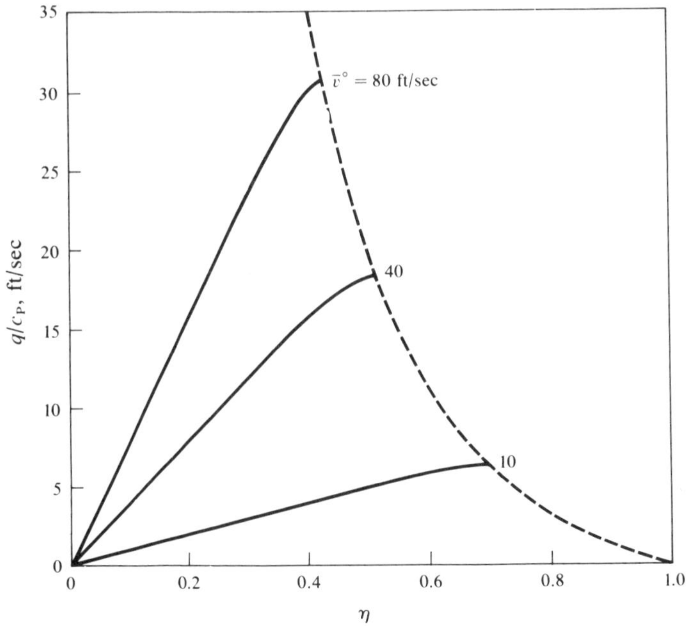

FIG. 4.6. Normalized flow q/c_P versus normalized concentration $\eta = c/c_P$ for the parabolic desired speed distribution function, Model 6 in Table 4.1. The solid curves represent the individual flow curves for various values of $\bar{v}^0$; dashed curve represents the collective flow curve. The probability of passing is taken as $P = 1-\eta$ and $c_P T = 0.1$ sec/ft.

range) desired speed distribution function is given in Fig. 4.7 for various normalized concentrations η. The normalization constant α associated with the irregular part is also given.

Calculations have been also made using a probability of passing of the form

$$P = 1-\eta^3 \tag{4.32}$$

instead of $P = 1-\eta$. As could be expected, the linear regime of the flow curve is extended, but no essentially new features have

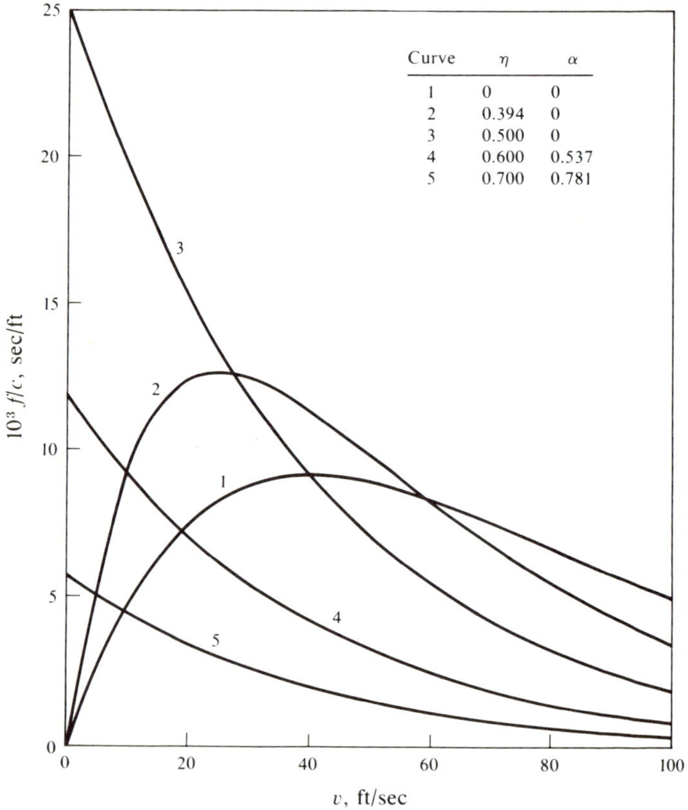

Curve	η	α
1	0	0
2	0.394	0
3	0.500	0
4	0.600	0.537
5	0.700	0.781

FIG. 4.7. Speed dependence for various concentrations, of the nonsingular part of the distribution function for the modified exponential, Model 2 in Table 4.1, an average desired speed $\bar{v}^0 = 80$ ft/sec, $c_P = 0.05$ car/ft, and $c_{PT} = 0.1$ sec/ft. The probability of passing is taken as $P = 1-\eta$. Curves 1, 2, and 3 correspond to individual flow states; curves 4 and 5 correspond to collective flow states.

been obtained. We theretore refer the reader to the original paper (Anderson *et al.*, 1962).

The results presented in this section show clearly the great variety of flow curves that may occur. There is in traffic flow theories no unifying feature comparable in simplicity and generality to the Maxwell velocity distribution law in kinetic flow theory in physics.

4.5 Transition between Individual and Collective Flow

First we show that if (see (4.14))

$$\left.\frac{f_0}{c}\right|_{v=0} \neq 0, \qquad \text{then } \alpha = 0 \text{ for all } 0 \leq c \leq c_P$$

assuming that f_0 is a continuous function of the velocity.

We write

$$f = \frac{f_0}{1+cT(1-P)(v-\bar{v})} + \alpha c\delta(v) \tag{4.33}$$

and let (see (3.12))

$$f_0 = c\tilde{f}_0 \tag{4.34}$$

where

$$\int_0^\infty dv\,\tilde{f}_0 = 1. \tag{4.35}$$

Then if we integrate (4.33) over all speeds, we obtain

$$c = c\int_0^\infty dv \frac{\tilde{f}_0}{1+cT(1-P)(v-\bar{v})} + \alpha c \tag{4.36}$$

or

$$\alpha = 1-\int_0^\infty dv \frac{\tilde{f}_0}{1+cT(1-P)(v-\bar{v})}. \tag{4.37}$$

We have already shown that

$$1-cT(1-P)\bar{v} > 0 \tag{4.38}$$

for $0 \leq c \leq c_{\text{crit}}$ in the individual flow regime with $\alpha \equiv 0$. Now we see that if

$$1-cT(1-P)\bar{v} = 0, \tag{4.39}$$

then

$$\alpha = 1-\int_0^\infty dv \frac{\tilde{f}_0}{cT(1-P)v}, \tag{4.40}$$

which is divergent for all f_0 such that

$$\tilde{f}_0(v = 0) \neq 0 \tag{4.41}$$

if $cT(1-P)$ is finite and non-zero. This condition on $cT(1-P)$ is valid for all $0 < c < c_P$ according to our assumptions regarding

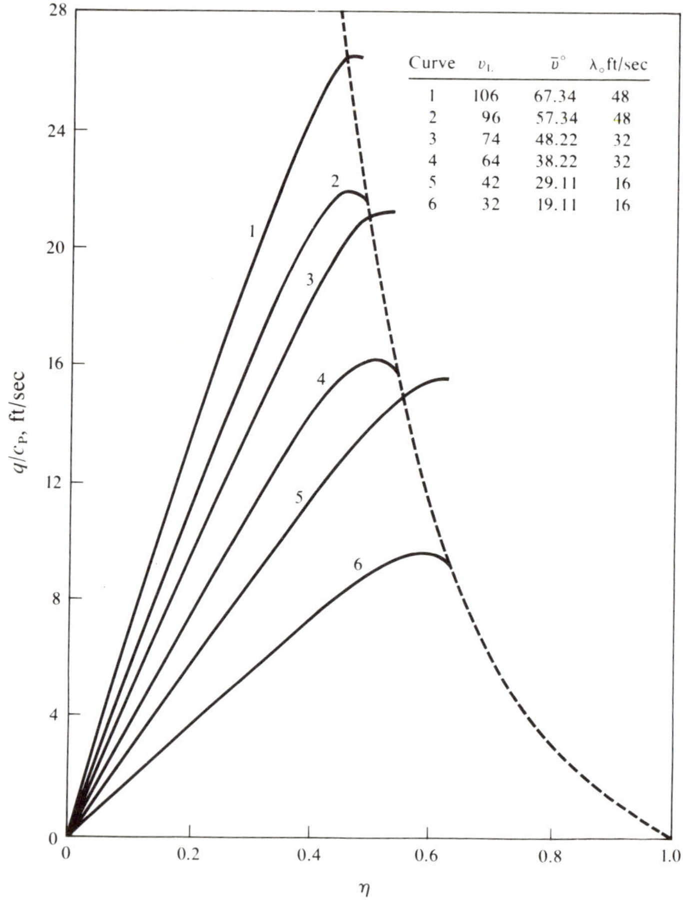

FIG. 4.8. Normalized flow q/c_P versus normalized concentration $\eta = c/c_P$ for the modified exponential finite speed range model. The desired speed distribution function

$$f_0 = cB(v_L - v)v \exp[-(v_L - v)/\lambda_0] \quad \text{for} \quad 0 \leq v \leq v_L,$$
$$= 0 \quad \text{for} \quad v_L \leq v,$$

where

$$B = \lambda_0^{-2}\{2v_L - (v_L + 2\lambda_0)[1 - \exp(-v_L/\lambda_0)]\}^{-1}.$$

the behavior of P and T. Hence, condition (4.40) cannot hold for any f_0 satisfying (4.41) in the range $0 < c < c_P$. Thus (4.39) cannot hold for any value of c in the range $0 \leq c < c_P$. Note that we pick up the equality sign because (4.39) obviously cannot hold for $c = 0$, which implies $\alpha \equiv 0$ for $0 < c < c_P$, as was to be shown.

Two cases are possible. Either the individual flow pattern meets the collective flow-concentration curve only at the point $\eta = 1$ (see Fig. 4.1) or the individual flow-concentration curves intersect the collective flow-concentration curve at various points for which $\eta < 1$ (see Fig. 4.8). It can be shown in the particular case of the exponential desired speed model that all the individual flow concentration curves intersect at the point $\eta = 1$. This can be done by noting that the determination of α corresponding to collective flow states leads to a divergence that can only be removed if $\eta = 1$.

We now consider more closely what happens at the intersection. Again two cases are possible. In the first case the individual solution cannot be continued after the intersection because the corresponding velocity distribution function becomes negative for small values of the velocity (see Section 4.2). Therefore, at the intersection we go directly from individual flow to collective flow with a *discontinuity in the slope* of the flow versus concentration curve. Or, the desired distribution function vanishes below $v = v_{\min}$. Then the individual flow curve can be continued beyond the intersection until a critical concentration is reached such that

$$1 + Tc(1-P)(v_{\min} - \bar{v}) = 0. \tag{4.42}$$

We have, therefore, the situation described schematically in Fig. 4.8. It should be noted that there exists now a region of concentrations in which our equation for the time-independent speed distribution function has two solutions. At the higher concentration

---

The solid curves that terminate on the dashed curve represent the individual flow concentration curves for various average desired speeds $\bar{v}^0$. The solid curves that extend beyond the dashed curve represent the individual flow-concentration curves for the case in which

$$f_0 = \begin{cases} 0, & 0 \leq v \leq v_{\min} \\ cB(v - v_{\min})\,(v_L - v)\exp[-(v_L - v)/\lambda_0], & 0 \leq v \leq v_L \end{cases}$$

where

$$B = \lambda_0^{-2}\{2\Delta v_{\max} - (\Delta v_{\max} + 2\lambda_0)\,[1 - \lambda^{-1}\exp(\Delta v_{\max})]\}^{-1}$$

and $\Delta v_{\max} = v_L - v_{\min}$. Note that $c_{PT} = 0.1$ sec/ft, $\lambda_0 = v_L/2$, and the probability of passing is taken as $P = 1 - \eta$. The dashed curve represents the collective flow-concentration curve.

we now have a finite jump of the flow from the individual to the collective flow curve corresponding to a discontinuity given by

$$\Delta\text{flow} = cv_{\text{min}} \tag{4.43}$$

where

$$f_0 = 0 \qquad \text{for} \qquad 0 \leqq v \leqq v_{\text{min}}.$$

It may be expected, however, that the part of the individual flow concentration curve at the right of the collective flow-concentration curve corresponds to a metastable regime, that is, a regime in which the flow is unstable with respect to fluctuations. The metastable flow states correspond to a higher average speed than the collective flow states at the same concentration. The effect of interactions is always to lower the average speed. This can be seen from the relationship between fluctuations in speed and average speed for time-independent situations. We have already established (see 3.28)

$$\bar{v} - \bar{v}^0 = Tc(1-P)[(\bar{v})^2 - \overline{v^2}] \leq 0.$$

We observe that the solution corresponding to a lower average speed also has the larger fluctuation. Therefore, we arrive at the intuitive conclusion: the stable flow pattern corresponds to the solution involving the largest fluctuations.

It should be pointed out, however, that our entire discussion of the collective flow was based on the assumption of the validity of a time-independent description. This, as we will see in the next chapter, is not a valid assumption, as the transition to collective flow and instabilities occur in the same range of concentration. In the quantitative description of the flow patterns at high concentrations, fluctuations will therefore play an essential role.

The prediction of a transition from individual to collective flow is certainly a nice feature of the basic kinetic equation (3.25). Indeed every driver experiences that at some moment, with increasing traffic concentration, his speed drops abruptly, and he is trapped in a collective flow to which he himself contributes.

As we observed in Section 4.2, there is some analogy with condensation phenomena. However, this analogy is more formal than physical. In the traffic case, the transition occurs as a result of the competition between relaxation and interference effects in the kinetic equation, and such effects do not exist in the ordinary Boltzmann equation nor in the physical theory of phase transitions.

CHAPTER 5

Statistical Hydrodynamics and Stability Theory

5.1 Local Steady State

In this chapter we follow closely the development in a paper by R. Balescu *et al.* (1967). We now return to a discussion of the kinetic equation in the general form (3.30)–(3.31). The time-independent homogeneous solution of the kinetic equation satisfies the nonlinear integral equation

$$O(f) = \mathscr{D}(f)+\mathscr{I}(f) = 0, \tag{5.1}$$

which contains the concentration c as a parameter. Let us write this solution in the form

$$f = F(c, v). \tag{5.2}$$

In the inhomogeneous case we may expect the same type of solution to remain valid if the time scale associated with the operators $\mathscr{D}$ and $\mathscr{I}$ is sufficiently short. This means that the characteristic times appearing in (5.1) are much smaller than the characteristic times associated with the inhomogeneity. Then the first approximation to the distribution function will be still given by (5.2) but *with the local value of the concentration*

$$f = F[c(x, t), v]. \tag{5.3}$$

This corresponds to what we called the local equilibrium solution in Chapter 3 (see (3.41)). An especially simple formulation of the condition of validity of this solution can be given if the sum of the functionals in (5.1) is written as

$$\mathscr{D}(f)+\mathscr{I}(f) = -\frac{f-F}{\tau_{\text{eff}}} \tag{5.4}$$

where τ_{eff} is an effective relaxation time. The local equilibrium solution can only be a good approximation whenever τ_{eff} may be considered negligible with respect to all other time scales involved.

Indeed then, the driver population always has the time necessary to relax to the distribution function f.

It should be noted that the steady-state solution (5.3) does not correspond to an equilibrium situation in the sense, for example, used in classical thermodynamics. For each of the two functionals we have

$$\mathscr{D}(F) \neq 0, \qquad \mathscr{I}(F) \neq 0, \tag{5.5}$$

with

$$\mathscr{D}(F) = -\mathscr{I}(F). \tag{5.6}$$

Equilibrium in the usual sense corresponds to the stronger requirement that each of the functionals vanish separately; that is,

$$\mathscr{D}(F) = \mathscr{I}(F) = 0. \tag{5.7}$$

This is only possible in the limit of low concentrations where F in (5.2) reduces to the desired velocity distribution

$$F \to f_0. \tag{5.8}$$

The nearest gas kinetic analogy to (5.6) would correspond to a situation in which the effect of the collisions is compensated by the effect of the external field or by the flow term in such a way that

$$\mathscr{C}(F) - \mathscr{F}\frac{\partial F}{\partial v} = 0. \tag{5.9}$$

Similarly, in the gas kinetic case we speak of equilibrium when the collision operator $\mathscr{C}(f)$ in (3.33) vanishes. In the more general case (5.9), we have a local steady state.

As we saw in Section 3.5, solutions of this type (5.3) lead directly to the kinematic waves of Lighthill and Whitham's traffic hydrodynamics. The fundamental question is to obtain a quantitative estimate of the validity of solutions of type (5.3), and this is one of the main objectives of the present chapter. A fluctuation is characterised either by some characteristic time τ_{inh} or by some characteristic length $L_{\text{inh}} \approx \tau_{\text{inh}}\bar{v}$, where $\bar{v}$ is of the order of the average speed of the traffic. We may always Fourier-analyze the perturbations. Therefore, using the language of Fourier analysis, we may expect that the local steady-state solution corresponds to a "long"-wavelength theory.

An unexpected and important result is that this condition about the scale of the perturbation is not sufficient, even if we consider local steady-state solutions. The short-wavelength behavior also plays an essential role, as it enters into the response of the system to perturbation. Indeed if we obtain growing perturbation waves, Eq. (5.3) will no longer correspond to the correct long-time behavior of the system. This is precisely the situation at high concentrations, as we will see later. We now briefly outline the method we will follow in the stability analysis.

5.2 Formulation of the Stability Problem

We again start with the generalized Boltzmann equation (3.30)

$$\frac{\partial f}{\partial t} + v\frac{\partial f}{\partial x} = O(f) \tag{5.10}$$

where $O(f)$ is the basic statistical traffic functional, which includes all the mechanisms that lead to a modification of the velocity distribution function of the cars. Next we take

$$f(x, v; t) = F(v) + \hat{f}(x, v; t) \tag{5.11}$$

and, similarly, for the concentration and flow

$$\begin{aligned} c(x; t) &= C + \hat{c}(x; t), \\ q(x; t) = c(x; t)\bar{v}(x; t) &= CU + \hat{q}(x; t), \end{aligned} \tag{5.12}$$

where the careted quantities represent small spatial and temporal perturbations. The functional F satisfies the equation

$$O(F) = 0, \tag{5.13}$$

which corresponds to a time- and space-independent solution of Eq. (5.1) appropriate to a given concentration C of cars. In Section 5.4 we present a method for the study of the distribution function f based on the special form (see 3.25) of $O(f)$ and appropriate to the case when f may be considered as a small perturbation.

We use an elementary Fourier–Laplace transform technique in order to determine an expression for the Fourier–Laplace transform $\nu_k(z)$ of $\hat{c}(x, t)$. Through the inverse Laplace transform, the behavior of $\hat{c}(x, t)$ will be determined by the nature and the location of the singularities of $\nu_k(z)$. Employing standard techniques, we find, in general, that the contributions to $\hat{c}(x, t)$ arise from the poles of $\nu_k(z)$ and the branch cuts associated with $\nu_k(z)$. We first discuss the qualitative features of the solution. The detailed solution will be found in Section 5.4.

The interesting feature is that we may express $\nu_k(z)$ as

$$\nu_k(z) = \frac{\Omega_k^{(1)}(z)+\Omega_k^{(2)}(z)}{\varepsilon_k(z)}, \tag{5.14}$$

where k is the wave number of the disturbance and z is a complex frequency. The function $\varepsilon_k(z)$ plays the essential role in our subsequent discussion. By analogy with similar terminology in physics (see Balescu, 1963), it may be called the *dispersion function.* An important point is that the singularities of $\varepsilon_k^{-1}(z)$ and $\Omega_k^{(2)}(z)$ are entirely determined by the form of the operator $O(f)$ and depend therefore only on the properties of the unperturbed traffic flow, whereas the singularities of $\Omega_k^{(1)}(z)$ depend on both the initial form of the perturbation and the parameters of the unperturbed traffic flow. Hence, there is a nontrivial separation of the investigation of stability of the traffic flow system into two areas of study. One is the study of a "characteristic" stability of the system, that is, contributions that arise from $\varepsilon_k(z)$ and $\Omega_k^{(2)}(z)$ and are determined only by the properties of the unperturbed traffic flow. These correspond to modes of the system that are excited by all perturbations of the system if the wavelength associated with the mode is present in the perturbation. The stability of the mode is independent of the nature of the perturbation. The second aspect is the study of a "perturbation-dependent" stability of the system, that is, contributions that arise from $\Omega_k^{(1)}(z)$ and are intimately related to the nature of the initial perturbation as well as depending on the parameters of the unperturbed traffic flow. Examples are considered later (see Sections 5.5 and 5.6).

The singularities of $\varepsilon_k^{-1}(z)$, $\Omega_k^{(1)}(z)$, and $\Omega_k^{(2)}(z)$ found in our calculation are poles and branch cuts. Each pole contributes a term of the form

$$(\text{amplitude})\cdot[\exp(-iz_0(k)t)]$$

where the amplitude is the residue at a pole, while a branch point z_b contributes damped terms of the asymptotic form $t^{-\alpha}\exp(-iz_bt)$ where $\alpha > 0$. The value of $z_0(k)$ may be separated into a real and an imaginary part

$$z_0(k) = \operatorname{Re} z_0(k)+i \operatorname{Im} z_0(k). \tag{5.15}$$

All positive values of $\text{Im } z_0(k)$ would mean amplification of initially present inhomogeneities. Therefore, the stability problem is reduced to the study of the sign of $\text{Im } z_0(k)$ corresponding to the poles and branch cuts. An important property to be studied is the behavior of $z_0(k)$ in the limit as $k \to 0$. This behavior determines the long-wavelength structure of the traffic flow.

There are two different possibilities: either

$$\lim_{k \to 0} \text{Im } z_0(k) \to 0, \tag{5.16}$$

or

$$\lim_{k \to 0} \text{Im } z_0(k) = \text{finite}. \tag{5.17}$$

In the first case, we have in the long-wavelength limit undamped wave propagation. As we will see, this behavior is characteristic of the singularities arising from the dispersion function.

On the contrary, the singularities that arise from $\Omega_k^{(2)}(z)$ behave as in (5.17) and have a finite absorption coefficient even for the long-wavelength limit. As a consequence, they are not related to the kinematic waves of continuum traffic hydrodynamics (see Chapter 2), which propagate without deformation in the long-wave limit.

We see that the kinetic equation leads to different types of waves. In low-temperature physics one would speak about zeroth, first, or second "sound." However, only one type of wave determined by the singularities of the dispersion function is related to kinematic waves. Before we proceed with the detailed theory, we present a qualitative discussion of the dispersion and the absorption of kinematic waves.

5.3 Dispersion and Absorption of Kinematic Waves

In this section we discuss only the contributions coming from the dispersion function $\varepsilon_k(z)$, which are the only ones related to the kinematic waves of continuum hydrodynamics. We write

$$z_0(k) = \omega(k) + i\gamma(k). \tag{5.18}$$

Then $\omega(k)$ leads to the dispersion of the kinematic waves and $\gamma(k)$ to their absorption.

In the long-wavelength limit, $k \to 0$, we find (see Section 5.7)

$$\omega(k) = v_g k + O(k^3). \tag{5.19}$$

The frequency versus k curve starts, therefore, with a linear range. The velocity v_g can be identified with the kinematic wave velocity V (see Section 2.1). It is positive for $C < C_m$ (where C_m is the concentration corresponding to the maximum of the flow-concentration curve), vanishes for $C = C_m$, and becomes negative for $C > C_m$. Moreover, we may also take into account the absorption coefficient $\gamma(k)$. Again in the long-wave limit

$$\gamma(k) = -Dk^2 + O(k^3). \tag{5.20}$$

For reasons that will become clear later (see Section 5.7) D may be called the diffusion coefficient. It should be noted that $\gamma(k)$ starts with a term in k^2 and not with a term in k. There exists, therefore (deferring the discussion of stability until later), a range of small wave vectors k for which we have the undamped kinematic waves of Lighthill and Whitham. These results, valid for long wavelengths, do not depend on specific assumptions about the desired speed distribution function.

Consider briefly the behavior of the diffusion coefficient as a function of concentration. For low concentrations we find

$$D \to 0 \qquad \text{for} \qquad C \to 0. \tag{5.21}$$

Moreover, not only does the diffusion coefficient vanish, but in general for $C \to 0$

$$\begin{aligned} \gamma &= 0, \\ \omega &= v_g k, \qquad \text{for all values of } k. \end{aligned} \tag{5.22}$$

In other words, in the limit of vanishingly small concentrations there is no absorption of perturbation waves and the macroscopic theory of kinematic waves becomes exact for all wavelengths. This is very natural and is a feature that is model independent. Indeed, the relaxation time T then goes to zero (see (4.5)). Note that we neglect all relaxation times that are not related to passing, so that at every moment the distribution function is maintained in the neighborhood of the desired speed distribution. No time delays and no irreversible processes appear at all.

Another remarkable feature is that the diffusion coefficient increases with concentration, goes through a maximum, vanishes for some concentration C_D, which in the model we treat in Section 5.8, is slightly smaller than C_{crit}, and then becomes negative. This change in sign indicates the beginning of the breakdown of the simple hydrodynamic theory based on a local theory steady-state solution, because beyond C_D perturbations of sufficiently long wavelengths are amplified instead of damped.

This behavior can be observed if we study the absorption coefficient $\gamma(k)$ as a function of k for different values of the concentration; it is studied in detail in Section 5.9 (see Fig. 5.4). For a very low concentration C_1, γ is negative for all wavelengths, and the perturbations are always damped; for C_2 we are below C_D and the damping is larger than for C_1; for C_5 we are just beyond C_D and perturbations of sufficiently long wavelengths become unstable. This reversal in the damping with increasing concentration can be completely accounted for in terms of the maximum of the diffusion coefficient (see the discussion in Section 5.8). For C_6 the domain of instability has grown; for C_7 the system is unstable with respect to all perturbations because the curve lies in the upper half-plane for all pertinent values of k. In the model we have studied in detail (see Section 5.5) instability with respect to all wavelengths occurs precisely at C_{crit}, that is, in the collective flow regime.

The occurrence of this instability probably indicates that the distribution function now becomes explicitly a function of time, and the unperturbed function F (see (5.11)) no longer gives a correct starting point. The macroscopic hydrodynamic theory is no longer valid. Broadly speaking, the distribution function is at some moment influenced more by the desires of the drivers and at the next moment by the interactions between the drivers. It would be most interesting to obtain explicit time-dependent solutions of the basic kinetic equation. This is, however, a problem of great mathematical complexity somewhat similar to the problem of fully developed turbulence in classical hydrodynamics.

5.4 Solution of the Linearized Kinetic Equation

We now present the exact solution of the kinetic equation (5.10) subject to the conditions (5.11)–(5.13), where we consider $\hat{f}(x, v; t)$ as a small perturbation of the steady-state solution $F(C, v)$, i.e., $f = F+\hat{f}$. We take the explicit form (3.25) of the kinetic equation

$$O(f) = (1-P)c(\bar{v}-v)f-\frac{f-f_0}{T}\,. \tag{5.23}$$

In order to linearize the Boltzmann equation (5.10), all quantities appearing in (5.10) and (5.23) must be expanded to first order in the perturbed quantities. In all subsequent mathematical developments careted quantities correspond to small spatial and temporal perturbations of the steady-state solution (5.13) and its moments. We find, as a consequence of the relations (5.11) and (5.12) and the assumption that T and P are functions of c only, that to first order in $\hat{c}$

$$
\begin{aligned}
T &= T_C + \left(\frac{\partial T}{\partial c}\right)_C \hat{c} = T_c + T_c' \hat{c}, \\
P &= P_C + \left(\frac{\partial P}{\partial c}\right)_C \hat{c} = P_c + P_c' \hat{c},
\end{aligned}
\tag{5.24}
$$

and because of the definition of the desired speed distribution function f_0 (i.e., the desired distribution of speeds is the same at all concentrations; see 3.13), we have

$$f_0 = (C+\hat{c})\mathsf{f}_0$$

and

$$\int_0^\infty dv\, \mathsf{f}_0(v) = 1. \tag{5.25}$$

Substituting relations (5.11)–(5.13) and (5.23)–(5.25) into the kinetic equation (5.10) yields

$$\frac{\partial f}{\partial t} + v\frac{\partial f}{\partial x} = -\frac{[F+\hat{f}-(C+\hat{c})\mathsf{f}_0]}{T_c(1+\hat{c}\,T_c'/Tc)} + (1-P_c-P_c'\hat{c})(CU+\hat{q}-Cv-\hat{c}v)(F+\hat{f}). \tag{5.26}$$

Equating like orders of the perturbed quantities, we obtain the zeroth-order equation

$$0 = -\frac{F-C\mathsf{f}_0}{T_c} + (1-P_c)C(U-v)F, \tag{5.27}$$

which is just a restatement of (5.13), and the first-order equation or linearized kinetic equation

$$\frac{\partial \hat{f}}{\partial t} + v\frac{\partial \hat{f}}{\partial x} = -\frac{(\hat{f}-\hat{c}\mathsf{f}_0)}{T_c} + (F-C\mathsf{f}_0)\frac{T_c'}{T_c^2}\hat{c} + (1-P_c)C(U-v)\hat{f} + [(1-P_c)(\hat{q}-\hat{c}v) - P_c'C(U-v)\hat{c}]F. \tag{5.28}$$

We now introduce the Fourier–Laplace transforms (the complex variable iz used here is usually denoted by $-s$ in the literature on Laplace transforms) corresponding to the various perturbed quantities

$$\begin{aligned}\hat{f}(x, v; t) &= \frac{1}{2\pi}\int dz\, e^{-izt}\int_{-\infty}^{+\infty} dk\, e^{ikx}\varrho_k(v; z),\\ \hat{c}(x; t) &= \frac{1}{2\pi}\int dz\, e^{-izt}\int_{-\infty}^{+\infty} dk\, e^{ikx}\nu_k(z),\\ \hat{q}(x; t) &= \frac{1}{2\pi}\int dz\, e^{-izt}\int_{-\infty}^{+\infty} dk\, e^{ikx}\mu_k(z),\end{aligned} \tag{5.29}$$

and specify the initial-value problem for the linearized kinetic equation (5.28) by assuming that

$$\hat{f}(x, v; 0) = \int_{-\infty}^{+\infty} dk\, e^{ikx}\hat{f}_k(v; 0). \tag{5.30}$$

Employing the relations (5.29) and (5.30), and noting that the Fourier–Laplace transform of $\partial\hat{f}/\partial t$ is

$$\mathscr{L}\left(\frac{\partial\hat{f}}{\partial t}\right) = -\frac{1}{2\pi}\int_{-\infty}^{+\infty} dk\, e^{ikx}\hat{f}_k(v; 0) - \frac{iz}{2\pi}\int_{-\infty}^{+\infty} dk\, e^{ikx}\varrho_k(v; z), \tag{5.31}$$

we obtain from the linearized kinetic equation (5.28)

$$\varrho_k(v; z) = \frac{\hat{f}_k + (1-P_c)F\mu_k - \{(1-P_c)U + [P_c' - (T_c'/T_c)(1-P_c)]C(U-v) - (T_cC)^{-1}\}F\nu_k}{[ik + (1-P_c)C]v - (1-P_c)CU - iz + T_c^{-1}}. \tag{5.32}$$

In obtaining Eq. (5.32) we have eliminated f_0 in terms of F using the zeroth-order equation (5.27).

Since

$$\int_0^\infty dv\, \varrho_k(v; z) = \nu_k(z) \tag{5.33}$$

and

$$\int_0^\infty dv\, v\varrho_k(v; z) = \mu_k(z),$$

a set of two simultaneous equations for $\nu_k(z)$ and $\mu_k(z)$ can be obtained by first integrating Eq. (5.32) over all speeds, then multiplying Eq. (5.32) by the speed v, and finally integrating over all speeds. We introduce the notation

$$\begin{pmatrix} A_n \\ \alpha_n \end{pmatrix} = \int_0^\infty dv\, v^n \frac{\begin{Bmatrix} \hat{f}_k(v; 0) \\ F(c, v) \end{Bmatrix}}{[ik + (1-P_c)C]v - (1-P_c)CU - iz + T_c^{-1}}. \tag{5.34}$$

There are simple recursion relations for the A_n's and α_n's that can be obtained by the method of partial fractions

$$\begin{Bmatrix} A_{n+1} \\ \alpha_{n+1} \end{Bmatrix} = \frac{1}{ik+(1-P_c)C}\left\{\int_0^\infty dv\, v^n \begin{Bmatrix} \hat{f}_k(v;0) \\ F(C,v) \end{Bmatrix} + [(1-P_c)CU+iz-T_c^{-1}]\begin{Bmatrix} A_n \\ \alpha_n \end{Bmatrix}\right\}. \tag{5.35}$$

Now an integration of Eq. (5.32) over all speeds yields

$$\begin{aligned} \nu_k(1+\{(1-P_c)U+[P_c'-(T_c'/T_c)(1-P_c)]CU-(CT_c)^{-1}\}\alpha_0 \\ -[P_c'-(T_c'/T_c)(1-P_c)]C\alpha_1)-\mu_k(1-P_c)\alpha_0 = A_0 \end{aligned} \tag{5.36}$$

and multiplication of Eq. (5.32) by a speed v followed by integration over all speeds yields

$$\begin{aligned} \nu_k(\{(1-P_c)U+[P_c'-(T_c'/T_c)(1-P_c)]CU-(CT_c)^{-1}\}\alpha_1 \\ -[P_c'-(T_c'/T_c)(1-P_c)]C\alpha_2)+\mu_k[1-(1-P_c)\alpha_1] = A_1. \end{aligned} \tag{5.37}$$

Thus we have obtained a set of two linear algebraic equations (5.36) and (5.37) for the two unknowns $\nu_k(z)$ and $\mu_k(z)$. Their determinant will be called the dispersion function and is denoted by $\varepsilon_k(z)$ (see also Section 5.2), which is given by

$$\begin{aligned} \varepsilon_k(z) = 1-\left[(1-P_c)\left(1-\frac{T_c'}{T_c}C\right)+P_c'C\right](\alpha_1-U\alpha_0)-\frac{\alpha_0}{CT_c} \\ +(1-P_c)\left[P_c'-\frac{T_c'}{T_c}(1-P_c)\right]C(\alpha_1^2-\alpha_0\alpha_2). \end{aligned} \tag{5.38}$$

The dispersion function contains α_0, α_1, and α_2. However, it may be reduced to the following linear form in α_0 by employing the method of partial fractions (see Eq. (5.35)):

$$\begin{aligned} \varepsilon_k(z) = 1-\left\{\frac{[(1-P_c)(1-CT_c'/T_c)+P_c'C]ik+(1-P_c)^2C}{[ik+(1-P_c)C]^2}\right\} \\ \times\left[C+\left(-\frac{1}{T_c}+iz-ikU\right)\alpha_0\right]-\frac{\alpha_0}{CT_c}. \end{aligned} \tag{5.39}$$

It is important to observe that the dispersion function (5.39) is determined through α_0 only by the properties of the unperturbed steady-state distribution function $F(C, v)$.

We find that the Fourier–Laplace transform $\nu_k(z)$ of the concentration perturbation $\hat{c}(z; t)$ is given by

$$\nu_k(z) = \frac{\Omega_k(z)}{\varepsilon_k(z)}, \tag{5.40}$$

where

$$\Omega_k(z) = \begin{vmatrix} A_0 & -(1-P_c)\alpha_0 \\ A_1 & 1-(1-P_c)\alpha_1 \end{vmatrix} = \Omega_k^{(1)}(z) + \Omega_k^{(2)}(z), \tag{5.41}$$

which can again be reduced by the method of partial fractions to

$$\Omega_k^{(1)}(z) = ikA_0, \qquad \Omega_k^{(2)}(z) = (1-P_c)\hat{c}_k\alpha_0, \tag{5.42}$$

where

$$\hat{c}(x; 0) = \int_{-\infty}^{+\infty} dk\, e^{ikx}\hat{c}_k. \tag{5.43}$$

We make the important observation here that the location of the singularities of $\varepsilon_k^{-1}(z)$ and $\Omega_k^{(2)}(z)$ is determined entirely by the properties of the unperturbed traffic flow, whereas the singularities of $\Omega_k^{(1)}(z)$ depend on both the initial perturbation and the parameters of the unperturbed traffic flow. The behavior of $\hat{c}(x; t)$ will be determined by the nature and location of these singularities; hence, we are led to the remarks made in Section 5.2 about the separation of the stability investigation into consideration of characteristic and perturbation-dependent stability.

5.5 Characteristic Stability of a Traffic Convoy

The characteristic stability of a traffic convoy is investigated in this section in order to illustrate the formalism developed in Section 5.4. For the purpose of this discussion, a traffic convoy is taken as a system of vehicle drivers in which all drivers desire to travel at the same speed, say U; that is,

$$f_0 = C\delta(v-U). \tag{5.44}$$

Hence, it follows from Eq. (5.23) that the steady-state solution is also given by

$$F(c, v) = C\delta(v-U). \tag{5.45}$$

Because of the simple form of the velocity distribution (5.44) and (5.45), we may determine the singularities corresponding to $\varepsilon_k(z)$, $\Omega_k^{(1)}(z)$, and $\Omega_k^{(2)}(z)$ for all values of the wavelength.

In this case the definition (5.34) allows us to write

$$\alpha_0 = \frac{C}{ikU - iz + T_c^{-1}} \tag{5.46}$$

and therefore the dispersion function (5.39) takes the simple form

$$\varepsilon_k(z) = 1 - \frac{\alpha_0}{CT_c} = 1 - \frac{1}{1 + ikUT_c - izT_c} \cdot \tag{5.47}$$

The zeros of $\varepsilon_k(z)$ correspond, therefore, to simple poles of $\nu_k(z)$, which are located at

$$z_0(k) = kU. \tag{5.48}$$

By comparison with (5.18)–(5.20), we see that in this case we have kinematic waves propagating without dispersion and absorption. The hydrodynamical theory is verified *exactly*. Moreover, the kinematic wave velocity coincides with the average velocity. This is in agreement with the general relation (2.4), as in this case the flow-concentration curve reduces to a constant as a consequence of (5.44).

Comparing (5.42) with (5.46), we see that the singularities of $\Omega_k^{(2)}$ are also given by simple poles, which are now related to the wavelength through

$$z_0(k) = kU - \frac{i}{T_c} \cdot \tag{5.49}$$

There is an essential difference between (5.48) and (5.49). In fact, (5.49) shows that the waves excited in the traffic medium through the singularities of $\Omega_k^{(2)}$ are damped in times of order T_c. The long-time behavior is determined entirely by the dispersion function. Therefore, we have obtained the interesting result that the analysis of the characteristic stability (independent of the nature of the initial perturbation) leads to contributions that are stable. Moreover, the differences between the case of $\Omega_k^{(2)}$ and $\varepsilon_k(z)$ are very clearly illustrated.

5.6 Comments on Perturbation-Dependent Stability

As we remarked in Section 5.4, by "perturbation-dependent" we mean those contributions that arise from the term $\Omega_k^{(1)}$ given by Eq. (5.42). These contributions, in addition to depending on the parameters of the unperturbed traffic flow, are intimately related to the nature of the initial perturbation. As a simple example, let us consider the case of an initial perturbation in the form of a velocity "spike" at the speed u, that is,

$$\hat{f}(x, v; 0) = \hat{c}(x; 0)\delta(v-u), \tag{5.50}$$

where $\delta(\xi)$ is the Dirac delta function.

Again the contributions associated with $\Omega_k^{(1)}(z)$ are simple pole contributions that are located at

$$z_0(k) = ku - i\left[(1-P)C(u-U) + \frac{1}{T}\right] \tag{5.51}$$

From our discussion of the transition between individual and collective flow we know that (see (4.42))

$$(1-P)C(v_{\min F} - U) + \frac{1}{T} \geq 0 \tag{5.52}$$

for all unperturbed concentrations $0 \leq C \leq C_P$, where $v_{\min F}$ is the minimum speed associated with the unperturbed flow, and the quantity C_P is the limiting concentration beyond which cars can no longer pass one another. It should also be noted that $v_{\min F} = v_{\min f_0}$, which is an expression of the fact that the lowest speed permitted in the unperturbed flow is the lowest desired speed. Hence, the poles given by (5.51) correspond to damped oscillations for all concentrations if

$$u \geq v_{\min F} \,. \tag{5.53}$$

Again we find that there is no dispersion and that there is a constant wavelength independent of absorption. In this example, however, instabilities become possible at sufficiently high concentrations. Indeed, instability may occur if

$$u < v_{\min F} \tag{5.54}$$

at a concentration that would depend on the difference $(v_{\min F} - u)$. The physical interpretation of this type of instability is that for sufficiently high concentrations, interactions may be successful in preventing the system from relaxing back to F, which is a distribution of higher speeds than $(F+f)$.

5.7 Diffusion Equation for Traffic Flow

We now return to the general theory of Section 5.4 and study in more detail the singularities due to the poles of the dispersion function in the limit of long wavelength. We are interested in the behavior of the zeros z_0 of the dispersion function (see (5.39))

$$\varepsilon_k(z) = 1 - \left\{\frac{aik+(1-P_c)^2C}{[ik+(1-P_c)C]^2}\right\}\left[C+\left(-\frac{1}{T_c}+iz-ik\bar{v}\right)\alpha_0\right]-\frac{\alpha_0}{CT_c} \quad (5.55)$$

in the neighborhood of $k = 0$. Note that we have used the abbreviation

$$a = (1-P_c)\left(1-\frac{T_c'}{T_c}\right)C+P_c'C$$

and that α_0 is given by Eq. (5.34). In order to obtain their behavior, we first note that

$$\varepsilon_k(z)|_{k=0} = 0 \quad (5.56)$$

implies that

$$z_0\alpha_0 = 0 \text{ for } k = 0. \quad (5.57)$$

Now it follows from physical arguments that α_0 is finite; hence, we conclude that

$$z_0(k = 0) = 0. \quad (5.58)$$

From continuity arguments and (5.58), we can expand $z_0(k)$ as a power series in k in the neighborhood of $k = 0$,

$$z_0(k) = \sum_{n=1}^{\infty} \left.\frac{\partial^n z_0}{\partial k^n}\right|_{k=0} k^n. \quad (5.59)$$

A determination of $z_0(k)$ can be obtained by substituting (5.59) into (5.55), rearranging this in a power series in k, and then invoking the condition that the coefficient k^n for all n must vanish identically. It is sufficient for our purposes to examine the linear and quadratic terms in this series.

From the coefficient of k we obtain

$$\begin{aligned} 0 = {} & \frac{i}{(1-P_c)^2C}[-2(1-P_c)+a]-i\left.\frac{\partial z_0}{\partial k}\right|_{k=0}\frac{T_c}{C}I_{1;\,0} \\ & +i\left\{\left[\frac{2(1-P_c)+a}{(1-P_c)^2C^2}\right]+U\frac{T_c}{C}\right\}I_{1;\,0} \end{aligned} \quad (5.60)$$

where we define

$$I_{n;\,\rho} = \int_0^\infty dv \frac{v^\rho F(C, v)}{[\Gamma(v+\beta)]^n}, \tag{5.61}$$

and

$$\Gamma = T_c(1-P_c)C, \tag{5.62}$$

$$\beta = \frac{1}{\Gamma} - U. \tag{5.63}$$

Remembering that (see (5.18))

$$z_0 = \omega + i\gamma$$

and setting the real and imaginary parts of (5.60) separately equal to zero, we find

$$v_g = \left.\frac{\partial\omega}{\partial k}\right|_{k=0} = U - \left[2 + \frac{a}{(1-P_c)}\right]\frac{1}{(1-P_c)CT_c}\left(1 - \frac{C}{I_{1;\,0}}\right), \tag{5.64}$$

and

$$\left.\frac{\partial\gamma}{\partial k}\right|_{k=0} = 0. \tag{5.65}$$

From the coefficient of k^2 we find that

$$\begin{aligned} 0 = {} & \frac{-2}{(1-P_c)^2C^2}\left[-3 + \frac{2a}{(1-P_c)}\right] - i\left.\frac{\partial^2 z_0}{\partial k^2}\right|_{k=0}\frac{T_c}{C}I_{1;\,0} \\ & + 2i\left(U - \left.\frac{\partial z_0}{\partial k}\right|_{k=0}\right)\left\{iT_c\left.\frac{\partial z}{\partial k}\right|_{k=0} I_{2;\,0}\right. \\ & \qquad \left. + i\frac{[-2(1-P_c)+a]}{(1-P_c)^2C^2}T_cI_{1;\,0} - i\frac{T^2}{C}I_{2;\,1}\right\} \\ & + \frac{[-3+2a/(1-P_c)]}{(1-P_c)^2C^2}\frac{2I_{1;\,0}}{C} - \frac{2i^2T_c}{(1-P_c)^2C^2} \\ & \qquad \times [-2(1-P_c)+a]I_{2;\,1}\,. \end{aligned} \tag{5.66}$$

Employing Eqs. (5.63) and (5.64) and setting the real and imaginary parts of (5.66) separately equal to zero, we obtain

$$\left.\frac{\partial^2\omega}{\partial k^2}\right|_{k=0} = 0, \tag{5.67}$$

and

$$D = -\frac{\partial^2\gamma}{\partial k^2}\bigg|_{k=0} = \left(\frac{C}{T_c} I_{1;\,0}\right)\left\{\frac{-2}{(1-P_c)C^2}\left[-3+\frac{2a}{(1-P_c)}\right]\left[1-\frac{I_{1;\,0}}{C}\right]\right.$$
$$+2(v_g-U)\left[\frac{T_c^2}{C}v_g I_{2;\,0}+\frac{[-2(1-P_c)+a]}{(1-P_c)^2C^2}T_c I_{1;\,0}-\frac{T_c^2}{C}I_{2;\,1}\right]$$
$$\left.+\frac{2T_c}{(1-P_c)^2C^2}[-2(1-P_c)+a]I_{2;\,1}\right\}. \tag{5.68}$$

Hence, we find that to lowest order in k

$$\begin{aligned} \operatorname{Re} z_0(k) &= \omega(k) = v_g k + O(k^3), \\ \operatorname{Im} z_0(k) &= \gamma(k) = -Dk^2 + O(k^3), \end{aligned} \tag{5.69}$$

in agreement with (5.19) and (5.20).

Now Eq. (5.69) implies that the long-wavelength contributions to $\hat{c}(t)$ from the poles of $\varepsilon_k^{-1}(z)$ satisfy the diffusion equation

$$\frac{\partial \hat{c}}{\partial t} + v_g\frac{\partial \hat{c}}{\partial x} = D\frac{\partial^2 \hat{c}}{\partial x^2}. \tag{5.70}$$

We may now write (see (5.29))

$$\begin{aligned} \hat{c}(x;t) &= \int_{-\infty}^{+\infty} dk\, e^{ikx}\, e^{iz_0(k)t}\nu(k) \\ &= \int_{-\infty}^{+\infty} dk\, e^{ikx}\, e^{-ikv_g t}\, e^{-Dk^2 t}\nu(k) \end{aligned} \tag{5.71}$$

and by straightforward differentiation see that this expression satisfies the partial differential equation (5.70). Since, according to (5.12),

$$c(x,t) = C + \hat{c}(x,t)$$

we note that C is independent of x and t, and that Eq. (5.70) can also be written in the form

$$\frac{\partial c}{\partial t} + v_g\frac{\partial c}{\partial x} = D\frac{\partial^2 c}{\partial x^2}. \tag{5.72}$$

Therefore, we obtain a diffusion equation for traffic, as shown by the form of Eq. (5.72); the coefficient D plays the same role as the diffusion coefficient in physics.

Our theory provides us with an explicit estimate of the diffusion coefficient. As shown by (5.68), the explicit expression for D is very complicated. An important feature that results from (5.66) is that (see (5.21))

$$D \to 0 \qquad \text{as} \qquad c \to 0.$$

This is a special case of (5.22), which, as we have seen, must be satisfied for all models. We now consider in greater detail the concentration dependence of the diffusion coefficient.

5.8 Concentration Dependence of the Diffusion Coefficient

All the results given in Section 5.7 were independent of the details of our model. However, in order to understand the concentration dependence of the long-wavelength behavior, we have considered in detail the simple case of a linear desired speed distribution function

$$f_0 = \begin{cases} \dfrac{2cv}{v_L^2}, & 0 \le v \le v_L \\ 0, & v_L < v. \end{cases} \tag{5.73}$$

This model has already been investigated in Section 4.4. For the purpose of numerical computation we set

$$v_L = 100 \text{ ft/sec}, \quad C_P = 0.05 \text{ car/ft}, \quad \tau = 2 \text{ sec}. \tag{5.74}$$

We now present numerical results for this model for the case of long wavelengths. Figures 5.1 and 5.2 show, respectively, the group velocity v_g and the diffusion coefficient D as a function of concentration. Examining Fig. 5.1, we see that v_g starts out at zero concentration with the value of the average desired speed $\bar{v}^0$ and then decreases as a function of concentration. At a concentration of 0.022 cars/ft, v_g changes sign; that is, just beyond this concentration the kinematic waves propagate in a direction opposite to the direction of flow for an observer stationed at the side of the roadway. The zero of v_g corresponds to the concentration at which the flow is a maximum, which is a general result.

Figure 5.2 reveals several interesting features. First, and one of the most significant results, is the fact that the diffusion coefficient becomes negative beyond $C = 0.021$ cars/ft; that is, the long-wavelength perturbations are not damped out in time. Thus, the steady-state solutions are unstable for long-wavelength perturbations beyond $C = 0.021$ cars/ft. Later in this chapter we will see that the long-wavelength perturbations are the first perturbations to result in instability. The steady-state solutions are stable to all wavelength perturbations for concentrations less than $C = 0.021$ cars/ft.

A second interesting feature is the existence of a maximum in the diffusion coefficient versus concentration curve. This is a result of the fact that, independent of the exact form for f_0, the diffusion coefficient vanishes in the limit of small concentration (see (5.23)) and becomes negative for sufficiently high concentrations.

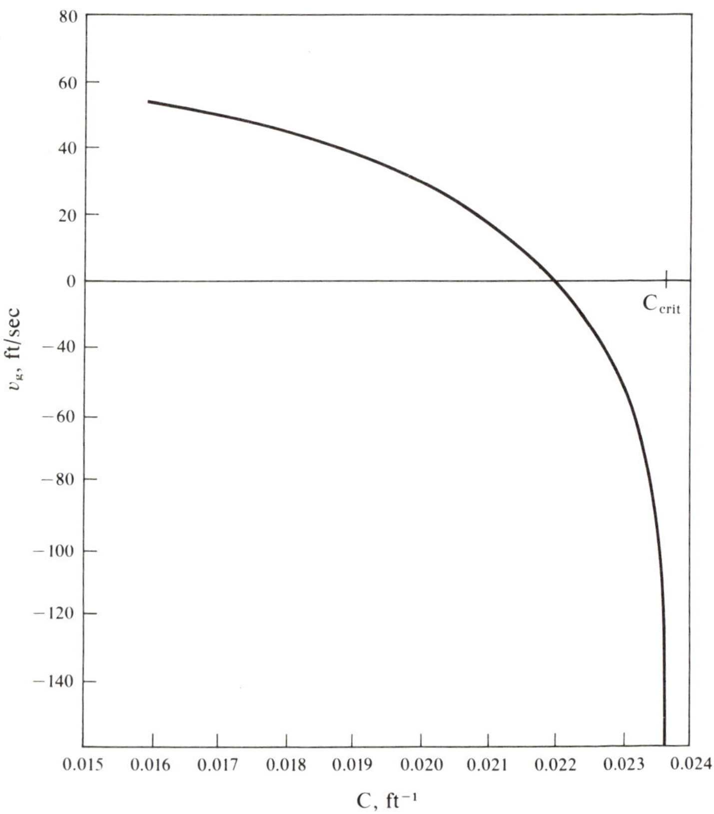

FIG. 5.1. The group velocity v_g versus concentration, C, for the linear desired speed distribution function, Model 5 in Table 4.1.

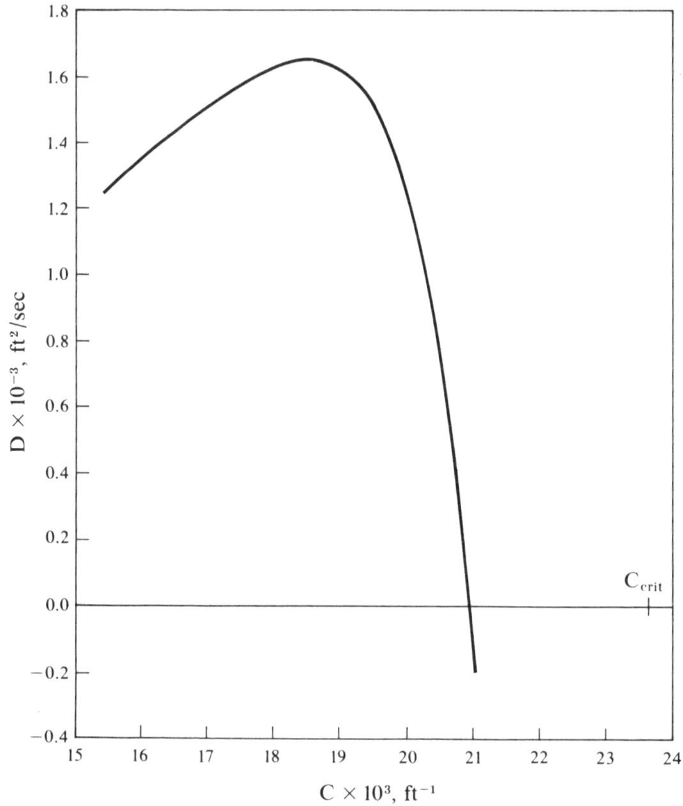

FIG. 5.2. The diffusion coefficient D versus the concentration for the linear desired speed distribution function, Model 5 in Table 4.1.

5.9 Short-Wavelength Behavior of Kinematic Waves

In this section we consider the whole range of relevant wavelengths. This range is determined by the requirement that the wavelength λ associated with the perturbation must be greater than or of the order of the average distance between cars $1/C$. Hence, since

$$k = \frac{2\pi}{\lambda}, \tag{5.75}$$

we find that in terms of the wave number k

$$0 \leq k \leq 2\pi C. \tag{5.76}$$

In section 5.7 we focused our attention on the case $k \sim 0$ and found some general results (i.e., results independent of f_0). To discuss that limiting case fully in Section 5.8 we had to turn to a specific model for f_0. Here we wish to consider the intermediate and short wavelengths and we again employ the model (5.73) for f_0.

Figure 5.3 shows $\omega \equiv \mathrm{Re}(z_0)$ as a function of the wave number k for various values of the concentration C. The expected feature

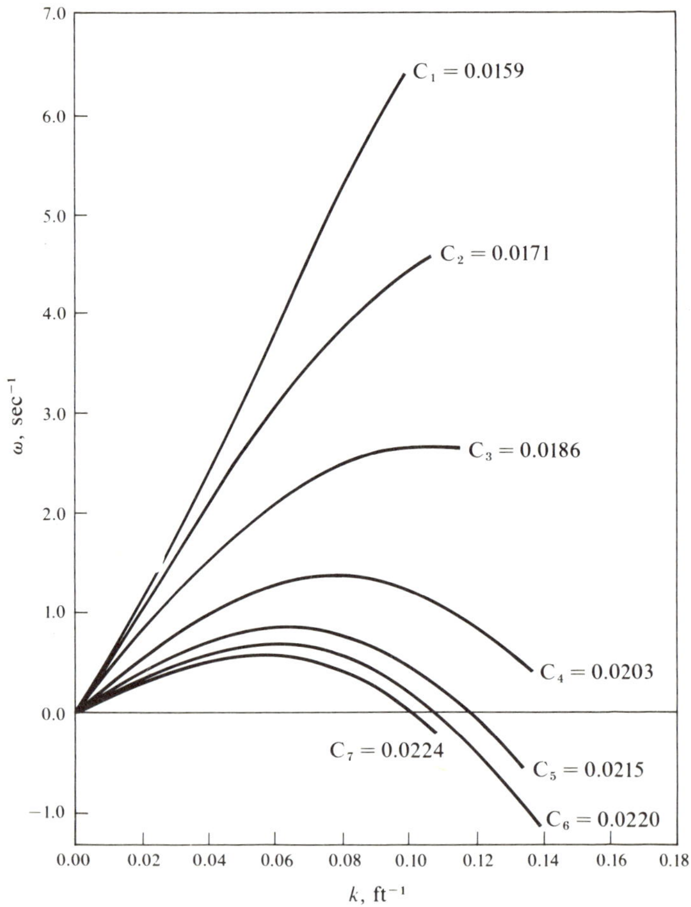

FIG. 5.3. The real part ω of the zero z_0 of the dispersion function $\varepsilon_k(z)$ versus the wave number k for various concentrations for the linear desired speed distribution function, Model 5 in Table 4.1.

here is the increase in dispersion for increasing concentration.

In Fig. 5.4 $\gamma \equiv \text{Im}(z_0)$ is plotted as a function of k for various concentrations C. We see, starting at $\gamma = 0$ for all k for $C = 0$ and proceeding to higher concentrations, that first the trend is to higher damping until we reach a concentration of $C \sim 0.0186$ cars/ft and then there is a reversal to smaller damping.

This reversal can be completely explained in terms of the maximum

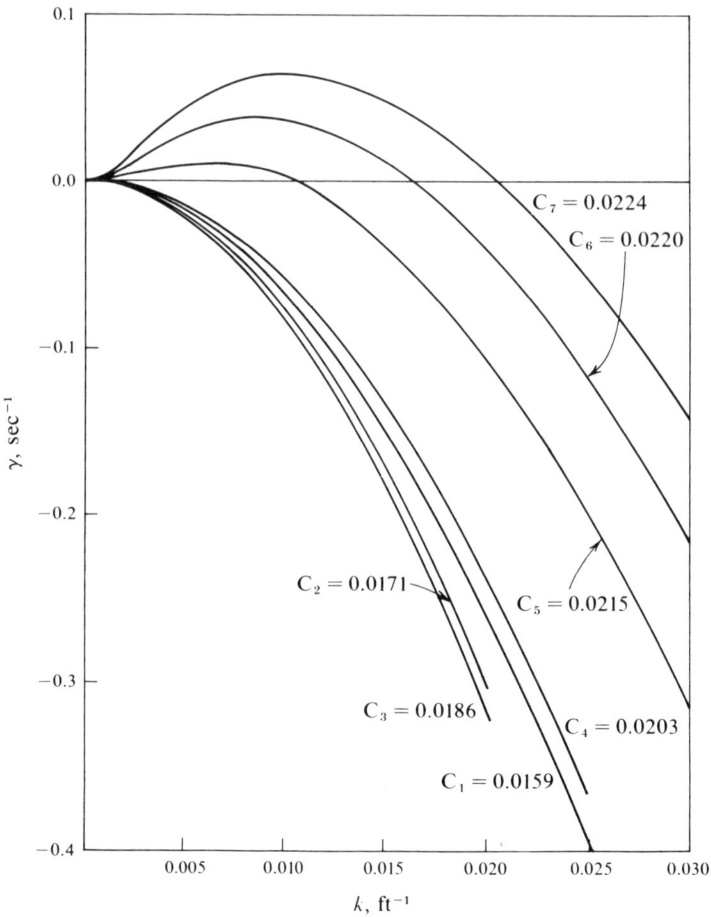

FIG. 5.4. The imaginary part of the zero z_0 of the dispersion function $\varepsilon_k(z)$ versus the wave number k for various concentrations for the linear desired speed distribution function, Model 5 in Table 4.1.

of the diffusion coefficient D and the long-wavelength behavior of γ. In particular, an examination of

$$\gamma = -Dk^2 \tag{5.77}$$

in the vicinity of D_{max} completely accounts for the reversal. As we proceed beyond $C_{D\max}$ the steady-state solutions are stable to all wavelengths until we reach $C = 0.021$ cars/ft. At this point we find that the long-wavelength perturbations begin to be unstable. It is important to note again that this is the point at which D begins to go negative. As we proceed to even higher concentrations, the solutions become unstable to a larger range of wavelengths. When we reach the critical value C_{crit}, we find that the solution is unstable to all wavelengths. Hence, an important result for this model is that the collective flow regime is completely unstable.

Figure 5.5 shows γ versus C for constant k and vividly exhibits the effect of D changing signs at $C \simeq 0.021$ cars/ft.

5.10 Contribution from Branch Cuts

We briefly study the contributions arising from $\Omega_k(z)$. Although we consider $\Omega_k^{(2)}(z)$, the same results apply to $\Omega_k^{(1)}(z)$ as well. Using (5.34) and (5.42) we have

$$\Omega_k^{(2)}(z) = (1-P_c)\hat{c}_k \int_0^\infty dv \frac{F(v, C)}{[ik+(1-P_c)C]v-(1-P_c)CU-iz+T_c^{-1}}. \tag{5.78}$$

An important special case corresponds to F equal to a constant in some interval v_{min} to v_{max} and zero outside. By the method of partial fractions (5.35) we come back to the type of expression where the function F is expressed as a polynomial in v.

This approximation leads us to

$$\begin{aligned}\Omega_k^{(2)}(z) = (1-P_c)\hat{c}_k C \\ [\ln\{[ik+(1-P_c)C]v_{max}-(1-P_c)CU-iz+T_c^{-1}\} \\ -\ln\{[ik+(1-P_c)C]v_{min}-(1-P_c)CU-iz+T_c^{-1}\}].\end{aligned} \tag{5.79}$$

We have two branch points associated with $z^{(1)}$ and $z^{(2)}$

$$\begin{aligned}z^{(1)} &= kv_{max}-i[(1-P_c)C(v_{max}-U)+T_c^{-1}],\\ z^{(2)} &= kv_{min}-i[(1-P_c)C(v_{min}-U)+T_c^{-1}].\end{aligned} \tag{5.80}$$

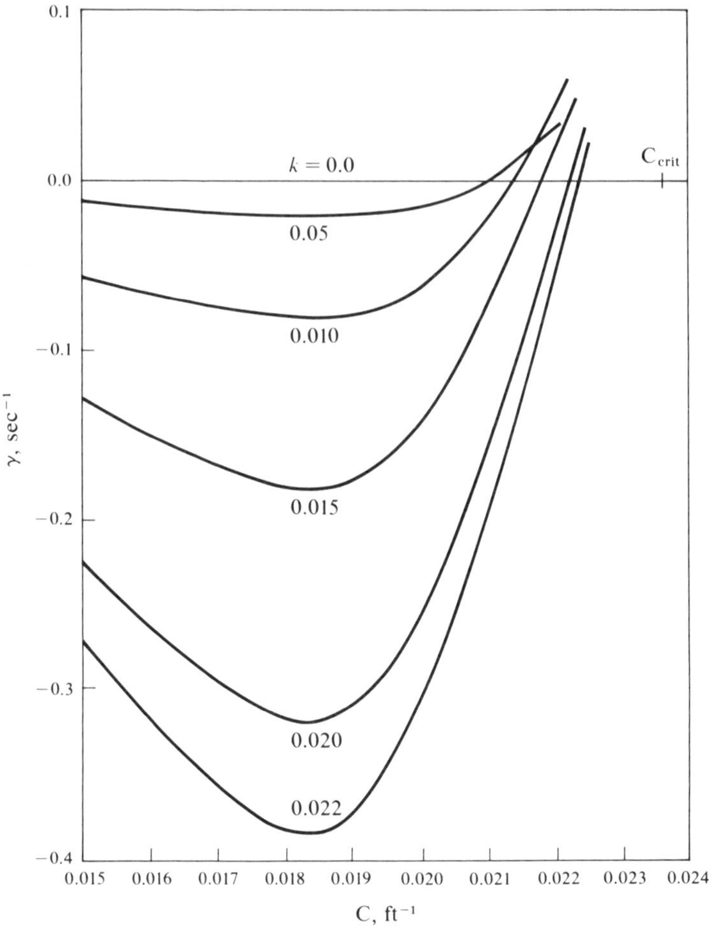

FIG. 5.5. The imaginary part γ of the zero z_0 of the dispersion function $\varepsilon_k(z)$ versus the concentration for various values of the wave number k for the linear desired speed distribution function, Model 5 in Table 4.1.

When we perform the z integration to go back to the time description, the branch points as usual have to be excluded by branch cuts. Note that in the individual flow regime the imaginary part of both $z^{(1)}$ and $z^{(2)}$ is negative.

We do not wish to go into a detailed discussion of the contributions of such branch cuts. It suffices for our purpose to present the following qualitative remarks.

1. These contributions are always model-dependent kinematic waves.

2. These contributions correspond to *finite damping*, even for $k \to 0$.

3. The distinction between kinematic waves and the other contributions becomes less clear for short wavelengths. Since we have not found any new qualitative features exhibited by the other types of waves (except the existence of perturbation-dependent stability, see Section 5.6), we will not go into more detail.

5.11 Limitations of the Local Steady-State Theory

Using the results presented in this chapter, we formulate in a semiquantitive way the limitations of the local steady-state theory (see Section 5.1), and therefore also of the hydrodynamic continuum theory (see Section 5.4).

a. Size of the Inhomogeneity

The wavelength of the inhomogeneity must be such that higher-order terms in k in (5.69) may be neglected. In the model considered in Section 5.8 this implies (see Fig. (5.3)) that, for instance,

$$k \leq 5.0 \times 10^{-2}\ \mathrm{ft}^{-1} \qquad \text{for} \qquad C \sim 0.2\ \text{cars/ft}.$$

For higher values of k we have to take into account the *dispersion* of kinematic waves. This is an alternate formulation of the condition

$$L_{\mathrm{inh}} \gg L_{\mathrm{eff}}$$

when (5.81)

$$L_{\mathrm{eff}} = \tau_{\mathrm{eff}} \bar{v}$$

and τ_{eff} is the characteristic time scale associated with $O(f)$.

b. Time Scale

The time must be sufficiently short so that the effect of the diffusion coefficient in (5.72) may be neglected. Therefore the local steady-state theory will lead to reliable results only if

$$t \ll \frac{L_{\mathrm{inh}}^2}{D}. \tag{5.82}$$

If the inequality (5.82) is not satisfied, the deformation by diffusion of kinematic waves, associated with a concentration inhomogeneity, becames important. For the example used in Section 5.8

$$D_{\max} \sim 10^3 \text{ ft}^2/\text{sec} \qquad \text{for} \qquad C \cong 0.02 \text{ car/ft}. \tag{5.83}$$

Therefore, the local steady-state theory is not in general a long-time theory, but is, in fact, a short-time theory for various concentration ranges and sizes of the inhomogeneity.

c. Instabilities

As we have seen, the local steady-state theory is based on an *exact* cancellation of the effect of the two operators $\mathscr{D}$ and $\mathscr{I}$ (see (5.6)). However, as we have seen for high concentrations of the order of those leading to collective flow, the corresponding solution is in general unstable, and a time-independent description breaks down.

This accounts qualitatively for the fact that beyond a limit of concentration $C \sim 50$ cars/mile, no steady flow is observed.

CHAPTER 6

Traffic Flow and Games of Strategy

6.1 Desired Speed Distribution Function and Games of Strategy

The theory of traffic flow presented in the previous chapters was based on the kinetic equation (3.30). More specifically, we have used the explicit form (3.25) of the kinetic equation. In this form of the theory, any "adjustment" of the drivers to their local environment was neglected. In Section 6.6, we considered the possibility of taking into account such adjustment effects by adding the supplementary term (3.45) to the kinetic equation. Now we would like to consider this adjustment problem from a different point of view, one closer to the theory of games of strategy as developed in the classical work of von Neumann and Morgenstern (1964).

We may consider driving on a highway as a kind of game in which the number of players is, however, very large; in fact, practically infinite. The rules of the game involve all quantities that appear in the functionals $\mathscr{D}(f)$ and $\mathscr{I}(f)$ (see (3.31)) such as f_0, T, and P. More precisely, the ideal distribution function f_0 characterizes the goals, or "program," that the driver population wishes to achieve. However, because of the conflicting aims of the drivers (as expressed by the existence of a finite dispersion in the function f_0) and their interaction, the effective distribution that is achieved is given by the solution f of the kinetic equation (3.25) and is, in general, different from f_0. It is only in the absence of interactions, that is, in circumstances corresponding to isolated Robinson Crusoe's, to use a terminology adopted in economic problems, that f and f_0 would coincide. In the presence of strong interactions, however, the strategy of the drivers may lead to traffic jams and instabilities. It may, therefore, become more favorable to introduce a new distribution function ϕ, which corresponds to the desired distribution function for the finite concentration that exists on the highway. This function must, of course, satisfy the condition

$$\phi(v) \to f_0(v) \qquad \text{for} \qquad c \to 0. \tag{6.1}$$

We now replace f_0 by ϕ in (3.25); the effective distribution function f is therefore a functional of ϕ. This replacement of f_0 by ϕ expresses the kind of adjustment introduced in our previous work through the addition of the term (3.45) to the kinetic equation.

In this perspective, traffic dynamics may be viewed as a successive feedback between a "preconceived" goal, represented by f_0 or by ϕ, and the adjustment necessary as a consequence of the competition between the drivers (Prigogine, 1965).

This may be represented schematically by the diagram

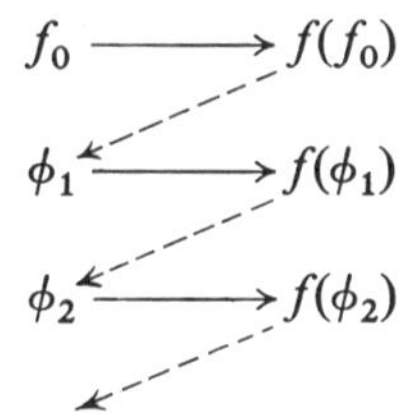

6.2 Rational Strategy

As an example we may consider the flow-concentration relationship for the modified exponential and the parabolic desired speed distribution functions. They were studied in detail in Section 4.4. It has been shown that there is a greater dispersion for a given average velocity $\bar{v}^0$ for the modified exponential than for the parabolic desired speed distribution function. In Fig. 6.1 we represent the flow-concentration relationships for the modified exponential desired speed distribution function with $\bar{v}^0 = 80$ ft/sec and the parabolic desired speed distribution function with $\bar{v}^0 = 60$ ft/sec.

As we saw in Section 4.3, deviations from the linear flow-concentration relationship are determined essentially by the dispersion of the desired speed distribution. It is, therefore, not surprising that the two curves in Fig. 6.1 cross. For low concentration the modified exponential speed distribution function is more favorable than the parabolic one, whereas for higher concentrations the situation is reversed. Then in the latter case an adjustment implying a shift to the parabolic desired speed distribution would increase the flow.

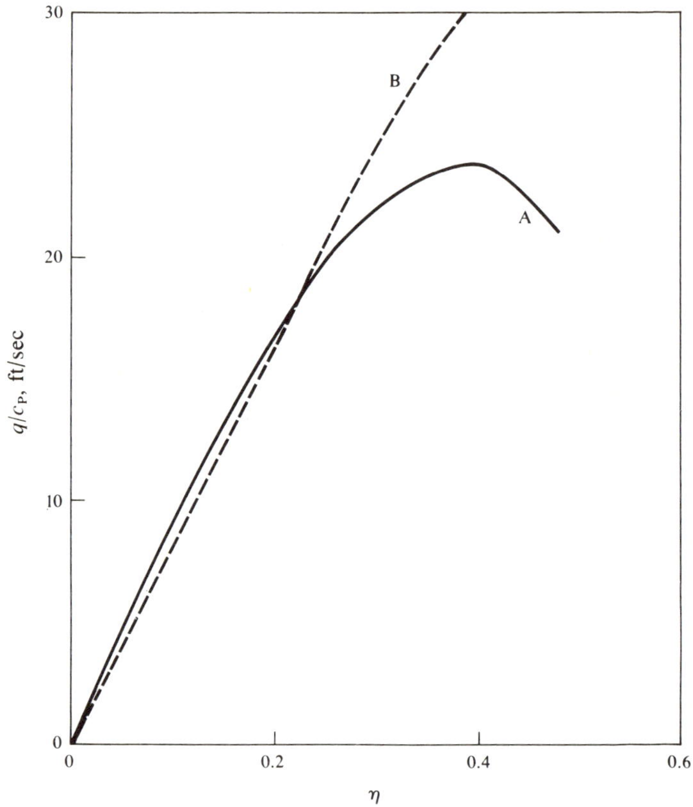

FIG. 6.1. The normalized flow q/c_P versus normalized concentration η, for (A) modified exponential speed distribution function, Model 2 (Table 4.1); and (B) parabolic Model 6 (Table 4.1).

We may now attempt to define the "rational" way of driving, restricting ourselves to time- and space-independent situations. Under these conditions it is natural to define the rational way of driving as the manner that corresponds to the highest possible flow for a given concentration. A similar definition is expressed by the requirement

$$I(c, \phi, f_0) = \int dv[f(\phi) - f_0]^2 = \text{minimum.} \tag{6.2}$$

This equality states that ϕ is chosen so as to make the difference between the actual distribution function f (which is a functional of ϕ) and the "original program" f_0 as small as possible. In this equation f will be determined, in time- and space-independent situations, by the solution of the kinetic equation (3.25), namely,

$$-\frac{f-\phi}{T}+c(1-P)(\bar{v}-v)f = 0. \tag{6.3}$$

It is clear that the two equations (6.2) and (6.3) form a closed system. In (6.2), for example, we may replace the function f by its value taken from (6.3), namely (see (4.8)),

$$f = \frac{\phi}{\lambda+\gamma v} \quad \text{with} \quad c = \int_0^\infty \frac{dv\phi}{\lambda_0+\gamma v}. \tag{6.4}$$

The minimum principle (6.2) then contains only the unknown function ϕ.

The main problem is to decide what is the class of admissible functions ϕ. If this class were unrestricted, it would always be possible to satisfy trivially the minimum problem by choosing

$$f(\phi) = f_0 . \tag{6.5}$$

However, this choice is in general not admissible. Such a choice implies that the average velocity corresponding to ϕ would be higher than that corresponding to f_0. But if such a distribution function were compatible with legal requirements, safety, and so on, then we would not understand why the population would have not chosen ϕ as the desired distribution function at low concentration. This leads us to require inequalities such as

$$\int dv\, vf_0 \geq \int dv\, v\phi, \tag{6.6}$$

which rule out the trivial solution (6.5). We next consider a simple example in detail.

6.3 Example of Rational Strategy and Adjustment

The problem of "rational driving" as defined in Section 6.2 corresponds to the minimization of the functional (6.2). In the interest of a qualitatively simple discussion, we write the solution (6.4) of the kinetic equation (6.3) in the explicit form

$$f_\phi = \phi+\tilde{T}c(\bar{v}^\phi-v)\phi \tag{6.7}$$

with

$$\tilde{T} = T(1-P). \tag{6.8}$$

As we have seen, this is the correct functional form only for low concentrations (see Section 4.3). Since this is not essential for our purpose, we will use Eq. (6.7) for arbitrary contributions.

As a special case of (6.7) we have

$$f_{f_0} = f_0 + \tilde{T}c(\bar{v}^{f_0} - v)f_0 \,. \tag{6.9}$$

Using (6.7) in (6.2), we obtain the explicit expression

$$I(c, \phi, f_0) = \int_0^\infty dv[(\phi - f_0) + \tilde{T}c(\bar{v}^{\phi} - v)\phi]^2. \tag{6.10}$$

Consider two extreme cases: In (6.9) we replace ϕ by f_0; that is, we make no adjustments whatsoever. We obtain

$$I(c, f_0, f_0) = \int_0^\infty dv(\tilde{T}c)^2(\bar{v}^{f_0} - v)^2 f_0^2 \,. \tag{6.11}$$

Clearly this expression depends primarily on the dispersion corresponding to f_0. If on the contrary we go to an extreme adjustment by replacing ϕ by a delta-like function in (6.10), the second term in the integrand of (6.10) disappears and we are left with

$$I(c, \delta, f_0) = \int_0^\infty dv(\delta - f_0)^2. \tag{6.12}$$

Note that in this case we consider the delta-like function as a sharply peaked continuous function, so that its square has a well-defined meaning. We expect that the ideal adjustment, which would make the integral I a minimum, lies somewhere between (6.11) and (6.12) and corresponds to a function that has a smaller dispersion than f_0, and therefore decreases the mutual hindering between the drivers without deforming the desired speed distribution function f_0 too much.

We consider the desired speed distribution

$$f_0 = \begin{cases} \dfrac{c}{2\Delta v} & \text{for} \quad v_0 - \Delta v < v < v_0 + \Delta v, \\ 0 & \text{otherwise.} \end{cases} \tag{6.13}$$

Therefore

$$\bar{v}^{f_0} = v_0. \tag{6.14}$$

We now consider the class of functions ϕ, of which (6.13) is a member, namely

$$\phi = \begin{cases} \dfrac{c}{2\Delta u} & \text{for} \quad v_0 - \Delta u < v < v_0 + \Delta u, \\ 0 & \text{otherwise,} \end{cases} \tag{6.15}$$

and write

$$\bar{v}^{\phi} = v_0. \tag{6.16}$$

The function ϕ therefore leads to the same average velocity but differs by the dispersion involved. In addition, we require that the admissible functions ϕ satisfy the subsidiary conditions

$$\Delta u \leq \Delta v. \tag{6.17}$$

The mutual hindrance between the drivers can only lead to an adjustment involving a decrease of the dispersion. This is the type of subsidiary condition that was mentioned in Section 6.2. The integral (6.10) can now be calculated explicitly as follows:

$$\begin{aligned} I(c, \phi, f_0) &= \frac{c^2}{4(\Delta u)^2} \int_{v_0 - \Delta u}^{v_0 + \Delta u} dv + \frac{c^2}{4(\Delta v)^2} \int_{v_0 - \Delta v}^{v_0 + \Delta v} dv \\ &\quad - 2\frac{c}{4\Delta u \Delta v} \int_{v_0 - \Delta u}^{v_0 + \Delta u} dv + \tilde{T}^2 c^2 \frac{c^2}{4(\Delta u)^2} \int_{v_0 - \Delta u}^{v_0 + \Delta u} dv (v - v_0)^2 \\ &= \frac{c^2}{2\Delta u} - \frac{c^2}{2\Delta v} + \frac{1}{6} \tilde{T}^2 c^4 (\Delta u). \end{aligned} \tag{6.18}$$

The only parameter that is open to variation is Δu. The minimum condition

$$\frac{\partial I}{\partial \Delta u} = 0 \tag{6.19}$$

leads to

$$\Delta u = \frac{\sqrt{3}}{\tilde{T} c} \tag{6.20}$$

and because of (6.17) this expression is valid for all concentrations larger than the critical concentration

$$c_{\text{crit}} = \frac{\sqrt{3}}{\tilde{T} \Delta v}. \tag{6.21}$$

Therefore we obtain the result represented schematically in Fig. 6.2. For the condition

$$c \leq c_{\text{crit}}, \tag{6.22}$$

the strategy function remains simply f_0; for larger concentrations, however, it is replaced by a function ϕ corresponding to a smaller dispersion.

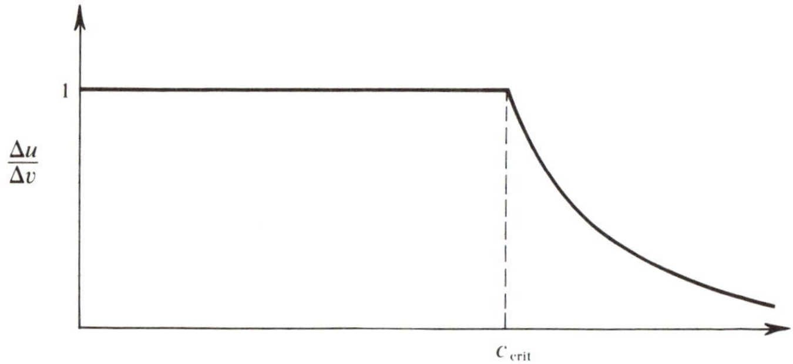

FIG. 6.2. Schematic representation of the reduction of the dispersion as a function of concentration.

It is interesting to verify explicitly that the adaptive behavior increases the average velocity and, therefore, the flow. Using (6.7) and (6.9) we have

$$\begin{aligned}\bar{v}^{f[f_0]} &= \bar{v}^{f_0} + \tilde{T}c[(\bar{v}^{f_0})^2 - \overline{v^2}^{f_0}]\\ &= v_0 + \tilde{T}c(v_0^2 - \overline{v^2}^{f_0}),\\ \bar{v}^{f[\phi]} &= \bar{v}^{\phi} + \tilde{T}c[(\bar{v}^{\phi})^2 - \overline{v^2}^{\phi}]\\ &= v_0 + \tilde{T}c(v_0^2 - \overline{v^2}^{\phi}).\end{aligned} \tag{6.23}$$

The notation $\bar{v}^{f[f_0]}$ means the average velocity calculated through the distribution function f, which itself is a functional of f_0; the meaning of $\bar{v}^{f[\phi]}$ is similar. Therefore (see 6.13), (6.15), and (6.17))

$$\begin{aligned}\Delta\bar{v} &= \bar{v}^{f[\phi]} - \bar{v}^{f[f_0]}\\ &= \tilde{T}c[\overline{v^2}^{f_0} - \overline{v^2}^{\phi}]\\ &= \frac{\tilde{T}c}{3}[(\Delta v)^2 - (\Delta u)^2] \geq 0.\end{aligned} \tag{6.24}$$

Using (6.20) and (6.21), we obtain an increase in the average velocity due to adaptive behavior

$$\Delta\bar{v} = \frac{1}{\tilde{T}(c_{\text{crit}})^2}[c^2 - (c_{\text{crit}})^2] \qquad \text{for} \qquad c \geq c_{\text{crit}}. \tag{6.25}$$

6.4 Adaptive Behavior and Collective Flow

It is, of course, possible to express the minimum condition (6.2) in an analytic form. As this requires lengthy, but elementary, transformations, we have given the calculations in the Appendix to this chapter.

An interesting observation is that the minimum condition is automatically satisfied in the collective region of flow. This is, indeed, very natural and means that then the flow pattern is independent of the choice of the strategy function f_0 or ϕ. This corresponds exactly to the very definition of the collective flow regime (see Section 4.2).

6.5 General Comments

The type of minimum principle we introduced in Section 6.2 may be applied to the determination of the "best" choice of other parameters that enter Eq. (6.3). We may, for example, consider f not only as a functional of ϕ, but also of the probability of passing $P(c)$. Until now we have always assumed (see (4.6)) that P is of the form

$$P = 1 - \frac{c}{c_p} \quad \text{for} \quad c < c_p \qquad \text{and} \qquad P = 0 \quad \text{for} \quad c > c_p. \tag{6.26}$$

While more general functions containing additional parameters could be used and these parameters fixed by the minimum condition (6.2), the main interest of the present approach most likely lies in its conceptional nature.

As already mentioned, we may consider driving a vehicle in the nature of a game played by a large number of participants. The rules of this game are expressed by all the quantities that enter into (6.3). The adjustment process described in the preceding sections may be considered as a change of these rules. We, therefore, have here an example of a game in which the rules are not fixed once for all, but vary according to circumstances. The modification of the rules implies a modification of behavior of the entire collectivity of participants. Finally, the process of adjustment implies a new kind of nonlinearity supplementary to the nonlinearity expressed by the interactions between the cars that was already introduced in the kinetic equation used in the preceding chapters.

6.6 Appendix: Analytic Formulation of the Variational Principle

Determine the strategy function $\phi(v)$ for a given concentration that makes the variance between the achieved distribution function f and the strategy function at zero concentration f_0 an extremum; that is (see Eq. (6.2))

$$\delta I = \delta\int_0^\infty dv(f-f_0)^2 = 0 \tag{6A.1}$$

where

$$\frac{\phi-f}{T(c)}+c[1-P(c)](\bar{v}-v)f = 0 \tag{6A.2}$$

and the concentration c is regarded as fixed.

The following are conditions imposed by the variational principle: Let

$$\phi(v,\ \varepsilon) = \phi(v,\ 0)+\varepsilon\eta(v) \tag{6A.3}$$

where

$$\int dv\ \phi(v,\ 0) = c \qquad \text{and} \qquad \int dv\ \eta(v) = 0; \tag{6A.4}$$

then

$$I(\varepsilon) = \int_0^\infty dv\left(\frac{\phi(v,\ \varepsilon)}{1+c(1-P)T\{v-\bar{v}[\phi(v,\ \varepsilon)]\}}-f_0\right)^2 \tag{6A.5}$$

where

$$\bar{v}[\phi(v,\ \varepsilon)] = \frac{1}{c}\int_0^\infty dv\ \frac{v\phi(v,\varepsilon)}{1+c(1-P)T\{v-\bar{v}[\phi(v,\ \varepsilon)]\}}. \tag{6A.6}$$

The condition for an extremum is

$$\left.\frac{\partial I(\varepsilon)}{\partial\varepsilon}\right|_{\varepsilon=0} = \int_0^\infty dv\ 2(f-f_0)\left[\frac{\eta(v)}{1+c(1-P)(v-\bar{v})}\right.$$
$$\left.+Tc(1-P)\left.\frac{\partial\bar{v}}{\partial\varepsilon}\right|_{\varepsilon=0}\frac{\phi(v,0)}{[1+c(1-P)T(v-\bar{v})]^2}\right] = 0 \tag{6A.7}$$

where

$$\bar{v} = \bar{v}[\phi(v,0)].$$

From Eq. (6A.6) we find that

$$\left.\frac{\partial\bar{v}}{\partial\varepsilon}\right|_{\varepsilon=0} = \frac{\dfrac{1}{c}\displaystyle\int_0^\infty dv\ \frac{v\eta(v)}{1+c(1-P)T(v-\bar{v})}}{1-(1-P)T\displaystyle\int_0^\infty dv\ \frac{v\phi(v,0)}{[1+c(1-P)T(v-\bar{v})]^2}}. \tag{6A.8}$$

A partial fraction decomposition and a definition will be useful in what follows, namely,

$$\int_0^\infty dv \frac{vg(v)}{1+c(1-P)T(v-\bar{v})}$$

$$= \frac{1}{c(1-P)T}\int_0^\infty dv\, g(v)\left[1-\frac{1-Tc(1-P)\bar{v}}{1+c(1-P)T(v-\bar{v})}\right] \tag{6A.9}$$

and

$$\mathscr{B}(v) = 1+c(1-P)T(v-\bar{v}). \tag{6A.10}$$

Employing Eqs. (6A.8)–(6A.10) and noting the cancellations that occur as a result of Eq. (6A.4), we obtain the following for Eq. (6A.7):

$$\left.\frac{\partial I}{\partial \varepsilon}\right|_{\varepsilon=0} = 2\int_0^\infty dv \int_0^\infty dv' \mathscr{H}\left\{\frac{\eta(v)}{\mathscr{B}(v)}\frac{\phi(v',0)}{[\mathscr{B}(v')]^2} - \frac{\phi(v,0)}{[\mathscr{B}(v)]^2}\frac{\eta(v')}{\mathscr{B}(v')}\right\} \tag{6A.11}$$

where

$$\mathscr{H} = \frac{[f(v)-f_0(v)]c^{-1}[1-c(1-P)T\bar{v}]}{1-(1-P)T\displaystyle\int_0^\infty dv' \frac{v'\phi(v',0)}{[\mathscr{B}(v')]^2}}. \tag{6A.12}$$

Equation (6A.11) vanishes identically for the collective flow since

$$1-c(1-P)T\bar{v} = 0 \tag{6A.13}$$

for $c > c_{\text{crit}}$ (see (4.10)). Since (6A.11) must vanish identically for arbitrary η subject to the normalization condition (6A.4), we conclude that for the individual flow regime

$$f = f_0 \tag{6A.14}$$

is always a possible solution, which is, however, generally incompatible with the subsidiary conditions that have to be introduced into the problem (see Sections 6.2 and 6.3).

Epilogue

We are well aware of the fragmentary character of this essay. Only highly idealized situations have been considered. Important problems such as town traffic and inhomogeneities due to entrance and exit effects have been completely ignored.

We hope that even in its present state our theory focuses attention on features that are likely to enter into any class of traffic problems. Every theory of traffic must eventually face the type of strategy problem introduced into our approach through the desired speed distribution. In the follow-the-leader approach this problem is implicit in the behavior of a lead driver. It is true that the existence of kinematic waves is independent of a detailed analysis of "strategy," but whenever we wish to calculate the properties of the kinematic waves such as speed or dispersion, we cannot avoid considering the effects of strategy. Even if we would like to describe the behavior of drivers in terms of some "transition probabilities" (e.g., for passing), these transition probabilities would presumably depend on the global situation and therefore on the overall concentration. In this case we would again face a strategy problem.

An obvious danger for a theory such as ours is that it may lead to a vicious circle. The behavior of drivers would be "explained" in terms of some desired distribution function, which in turn is introduced to describe the behavior of the drivers! This is similar to the well-known difficulty in economic theory, where the "utility" concept may easily lead to meaningless statements (see Samuelson, 1966).

At present, in the absence of adequate experimental data we may only hope that the desired speed distribution function has sufficient "stability" so that it may be extrapolated to predict the flow patterns that exist under different conditions. The validity of this approach can only be judged by its empirical success. Even so, we hope to have clearly formulated the problem and developed the theory to a point where the meeting of theoretical concepts with experimental observations will be fruitful.

References

Almond, Joyce (Ed.) (1965). *Proc. Second International Symposium on the Theory of Traffic Flow*, The Organization for Economic Co-operation and Development, Paris.

Anderson, R. L., Herman, R., and Prigogine, I. (1962). On the Statistical Distribution Function Theory of Traffic Flow, *Opns. Res.*, **10**, 180–196.

Ashton, W. D. (1966). *The Theory of Road Traffic Flow*, Methuen, London.

Australian Road Research Board Proc. (1962, 1964, 1966, 1968). Ramsay, Ware Publ. Pty. Ltd.

Balescu, R. (1963). *Statistical Mechanics of Charged Particles*, Wiley (Interscience), New York.

Balescu, R., Prigogine, I., Anderson, R. L., and Herman, R. (1967). Statistical Hydrodynamics of Traffic Flow, *Vehicular Traffic Science* [*Proc. Third International Symposium on the Theory of Traffic Flow*, L. C. Edie *et al.* (Eds.)], p. 72, American Elsevier, New York.

Bierley, R. L. (1963). Investigation of an Intervehicle Spacing Display, *Highway Res. Board Record*, No. 25, p. 58.

Borel, E. (1925). *Méchanique Statistique*, Gauthier-Villars, Paris.

Brainard, N. R., Trabold, W. G., and Becker, F. C. (1961). A Digital Data Recording System for Traffic Flow Analysis, *Proc. National Electronics Conference*, **17**, 435.

Brout, R., and Prigogine, I. (1956). Statistical Mechanics of Irreversible Processes, P. VIII: General Theory of Weakly Coupled Systems, *Physica*, **22**, 621.

Chandler, R. E., Herman, R., and Montroll, E. W. (1958). Traffic Dynamics: Study in Car Following, *Opns. Res.*, **6**, 165–184.

Edie, L. C. (1961). Car Following and Steady State Theory for Non-Congested Traffic, *Opns. Res.*, **9**, 66.

Edie, L. C., Foote, R. S., Herman, R., and Rothery, R. (1963). Analysis of Single-Lane Traffic Flow, *Traff. Engng.*, **33**, 21.

Edie, L. C., Herman, R., and Rothery, R. (Eds.) (1967). *Vehicular Traffic Science* (*Proc. Third International Symposium on the Theory of Traffic Flow*), American Elsevier, New York.

Franklin, R. E. (1965). On the Flow Concentration Relationship for Traffic, *Proc. Second International Symposium on the Theory of Traffic Flow*, The Organization for Economic Co-operation and Development, Paris.

Gafarian, A. V., Munjal, P. K., and Pahl, J. (1970). An Experimental Validation of the Boltzmann-Type Statistical Models for Multi-Lane Traffic Flow, Systems Development Corporation, Santa Monica, California; *Transportation Res.*, **5**, in press.

Gazis, D. C. (1967). Mathematical Theory of Automobile Traffic, *Science*, **157**, 273.

Gazis, D. C., Herman, R., and Potts, R. B. (1959). Car Following Theory of Steady-State Flow, *Opns. Res.*, **7**, 499.

Gazis, D. C., Herman, R., and Rothery, R. (1961). Non-linear Follow-the-Leader Models of Traffic Flow, *Opns. Res.*, **9**, 545.

Gerlough, D. L., and Capelle, D. G. (Eds.) (1964). An Introduction to Traffic Flow Theory, *Highway Res. Board Special Rep.*, **79**.

Greenberg, H. (1959). An Analysis of Traffic Flow, *Opns. Res.*, **7**, 79.

Haight, F. A. (1963). *Mathematical Theories of Traffic Flow*, Academic Press, New York.

Herman, R. (Ed.) (1961). *Theory of Traffic Flow* (*Proc. First International Symposium on the Theory of Traffic Flow*), Elsevier, Amsterdam.

Herman, R. (1966). Theoretical Research & Experimental Studies in Vehicular Traffic, *Proc. Australian Road Res. Board*, **3**, 25.

Herman, R., and Gardels, K. (1963). Vehicular Traffic Flow, *Scientific American*, **209**, No. 6, 3.

Herman, R., and Weiss, G. (1961). Comments on the Highway Crossing Problem, *Opns. Res.*, **9**, 828.

Herman, R., Montroll, E. W., Potts, R. B., and Rothery, R. (1959). Traffic Dynamics: Analysis of Stability in Car-Following, *Opns. Res.*, **7**, 86.

Hirschfelder, J. O., Curtiss, C. F., and Bird, R. B. (1964). *Molecular Theory of Gases and Liquids*, Wiley, New York.

Horton, T. R. (Ed.) (1965). *Traffic Control Theory and Instrumentation*, Plenum Press, New York.

Kraus, J. L. (1966). Distribution of Vehicle Speeds, unpublished manuscript, The Port of New York Authority.

Landau, L. D., and Lifshitz, E. M. (1959). *Fluid Mechanics*, Addison-Wesley, Reading, Massachusetts.

Leutzbach, W., and Baron, P. (Eds) (1969). Beiträge zur Theorie des Verkehrsflusses, (*Proc. Fourth International Symposium on the Theory of Traffic Flow*), Herausgegeben vom Bundesminster für Verkehr, Bonn.

Leutzbach, W., and Egert, P. (1965). Overtaking in Free Traffic, *Proc. Second International Symposium on the Theory of Traffic Flow*, The Organization for Economic Co-operation and Development, Paris.

Lighthill, M. J., and Whitham, G. B. (1955). On Kinematic Waves: II. A Theory of Traffic on Long Crowded Roads, *Proc. Roy. Soc.*, **A229**, 318.

Lowe, Adolph (1965). *On Economic Knowledge*, Harper & Row, New York.

Miller, A. J. (1961a). A Queueing Model for Road Traffic Flow, *J. Roy. Statist. Soc.*, **B23**, 64.

Miller, A. J. (1961b). Traffic Flow Treated as a Stochastic Process, *Theory of Traffic Flow* [*Proc. First International Symposium on the Theory of Traffic Flow*, R. Herman (Ed.)], Elsevier, Amsterdam.

Miller, A. J. (1962). Road Traffic Flow Considered as a Stochastic Process, *Proc. Cambridge Phil. Soc.*, **58**, 312.

Miller, A. J. (1963). Analysis of Bunching in Rural Two-Lane Traffic, *Opns. Res.*, **11**, 236.

Miller, A. J. (1967). Queueing in Rural Traffic, *Vehicular Traffic Science* [*Proc. Third International Symposium on the Theory of Traffic Flow*, L. C. Edie, R. Herman, and R. Rothery (Eds.)], American Elsevier, New York.

Munjal, P., and Pahl J. (1969). An Analysis of the Boltzmann-Type Statistical Models for Multi-Lane Traffic Flow, *Transportation Res.*, **3**, 151.

Pipes, L. A. (1953). An Operational Analysis of Traffic Dynamics, *J. Appl. Phys.*, **24**, 274.

Prigogine, I. (1961). A Boltzmann-like Approach to the Statistical Theory of Traffic Flow, presented at the First International Symposium on the Theory of Traffic Flow (1959) and published in *Theory of Traffic Flow*, R. Herman (Ed.), Elsevier, Amsterdam.

Prigogine, I. (1965). Sur la Théorie Mathématique de Trafic Véhiculaire, *Cahiers du Centre d'Etudes de Recherche Opérationnelle*, **7**, 177.

Prigogine, I., and Andrews, F. C. (1960). A Boltzmann-like Approach for Traffic Flow, *Opns. Res.*, **8**, 789.

Prigogine, I., and Herman, R. (1970). Comments on "An Analysis of the Boltzmann-type Statistical Models for Multi-Lane Traffic Flow," *Transportation Res.*, **4**, 113.

Prigogine, I., Herman, R., and Anderson, R. L. (1962a). On Individual and Collective Flow, *Bull. Acad. Royale de Belgique*, **48**, 792.

Prigogine, I., Resibois, P., Herman, R., and Anderson, R. L. (1962b). On a Generalized Boltzmann-like Approach for Traffic Flow, *Bull. Acad. Royale de Belgique*, **48**, 805.

Prigogine, I., Anderson, R. L., and Herman, R. (1967). Local Steady State Theory and Macroscopic Hydrodynamics of Traffic Flow, *Vehicular Traffic Science* [*Proc. Third International Symposium on the Theory of Traffic Flow*, L. C. Edie *et al.* (Eds.)], American Elsevier, New York.

Reuschel, A. (1950). Fahrzeubgewegungen in der Kolonne bei Gleichförmig Beschleunigtem oder Verzögertem Leitfahrzeug, *Z. Osterr. Ingr.-Arch. Vereins*, **95**, 59, 1950; *Osterr. Ingr.-Arch.*, **4**, 193.

Richards, P. I. (1956). Shock Waves on the Highway, *Opns. Res.*, **4**, 42.

Road Research Laboratory, (1965). *Research on Road Traffic*, Her Majesty's Stationery Office, London.

Samuelson, Paul A. (1966). *Foundations of Economic Analysis*, Harvard Univ. Press, Cambridge, Massachusetts.

Titmarsh, E. C. (1932). *Introduction to the Theory of Fourier Integrals*, Oxford (Clarendon Press), London and New York.

Tolman, R. C. (1938). *The Principles of Statistical Mechanics*, Oxford (Clarendon Press), London and New York.

Von Neumann, J., and Morgenstern, O. (1964). *Theory of Games and Economic Behavior*, Wiley, New York.

Weiss, G. H. (1965). A Survey of Some Recent Research in Road Traffic, *Congestion Theory*, W. L. Smith and W. Wilkins (Eds.), p. 253, Univ. of North Carolina Press, Chapel Hill.

Weiss, G., and Herman, R. (1962). Statistical Properties of Low-Density Traffic, *Quart. Appl. Math.*, **20**, 121.

AUTHOR INDEX

Numbers set in *italics* designate page numbers on which complete literature citations are given.

SUBJECT INDEX

This book is set in Times Roman, a monotype face and one of the most useful of the contemporary body types. It was designed by Stanley Morrison for use in *The Times* of London.

The type was set in Great Britain by Willmer Brothers Limited, Birkenhead. The book was printed by offset lithography at Rae Publishing Company, Inc., Cedar Grove, New Jersey, and bound by Publishers Book Bindery, Inc., New York. The paper is P&S Glatfelter Old Forge FFF and the bookcloth was manufactured by The Holliston Mills, Inc.